HOLT

Biology

Note-taking Workbook

HOLT, RINEHART AND WINSTON
A Harcourt Education Company
Orlando • Austin • New York • San Diego • London

Printed in the United States of America.

ISBN-13: 978-0-03-093221-2
ISBN-10: 0-03-093221-1

4 5 6 7 862 10 09 08

Contents

Contents

Name ______________ Class ______________ Date ______________

Note-taking Workbook

Biology and You

Section: The Nature of Science

SCIENTIFIC THOUGHT

Key Idea: Scientific thought involves ______________ using evidence to ______________ being ______________ about ideas and ______________

skepticism is ______________

Additional notes about Questioning Ideas: ______________

Additional notes about Discovery and Change: ______________

UNIVERSAL LAWS

Key Idea: Science is governed by ______________

These truths are called ______________.

Additional notes about Universal Laws: ______________

The word **aspect** means ______________

SCIENCE AND ETHICS

Key Idea: Because ____________________ and discovery can have serious ____________________, scientific investigations require ____________________.

Additional notes about Science and Ethics: ____________________

Reading Check: Why is it important that scientific investigations be done ethically? ____________________

WHY STUDY SCIENCE?

Key Idea: An understanding of science can help you ____________________, be a ____________________, and become a ____________________.

Additional notes about Why Study Science?: ____________________

In your own words, write the everyday meaning for the word law.

____

How does the everyday meaning compare to the science meaning of law?

____

Note-taking Workbook

Biology and You

Section: Scientific Methods

BEGINNING A SCIENTIFIC INVESTIGATION

Key Idea: Most scientific investigations begin with ______________ that lead to ______________.

Observation is ______________________________

A **hypothesis** is ______________________________

Additional notes about Making Observations: ______________________________

Additional notes about Formulating a Hypothesis: ______________________________

SCIENTIFIC EXPERIMENTS

Key Idea: Scientists conduct ______________ or perform ______________ in order to test a ______________.

An **experiment** is a ______________________________

A **control group** serves as a ______________________________

Additional notes about Controlled Experiments: ______________________________

Additional notes about Study Without Experimentation: ______________

____

____

Additional notes about Analyzing Results: ______________

____

____

Additional notes about Drawing Conclusions and Verifying Results: ______

____

____

Additional notes about Considering Bias: ______________

____

____

List the hypotheses that were proposed to explain the decrease in Canada geese in Chicago. ______________

____

Which hypothesis was supported by evidence? ______________

____

SCIENTIFIC THEORIES

Key Idea: The main difference between a ______________ and a hypothesis is that a hypothesis is a ______________ and a theory ______________.

A **theory** is a ______________

____

The steps involved in developing a theory are:

1. Questions
2. ____
3. Experimentation
4. ____
5. ____
6. Theory

Reading Check: How does the scientific use of the word theory differ from how it is used by the general public? ______________________________

____

____

Additional notes about Constructing a Theory: ______________________________

____

____

Name ____________________ Class ______________ Date ______________

Note-taking Workbook

Biology and You

Section: Tools and Techniques

MEASUREMENT SYSTEMS

Key Idea: The ____________ is used by all scientists because ____________. SI is also preferred by scientists because it is ______________________________

SI is ______________________________

Reading Check: How are prefixes used in names of SI units? ______________________________

Additional notes about Measurement Systems: ______________________________

LAB TECHNIQUES

Key Idea: In the lab, scientists always keep ____________ and perform ____________. Many scientists also use ____________ such as ____________ and specialized procedures such as ____________.

A **technique** is ______________________________

Additional notes about Microscopy: ______________________________

Additional notes about Sterile Technique: ______________________________

Additional notes about Collecting Data Remotely: ____________________

____

____

Reading Check: When might sterile technique be used in a lab?

____

SAFETY

Key Idea: Scientists must use caution when ______________ or doing ______________ to avoid dangers such as ____________________

____

Some guidelines for working safely in a lab are: ____________________

____

____

____

Complete the idea wheel that you started at the beginning of the chapter:

Tools and Techniques

Additional notes about Safety: ____________________

____

____

Name ______________________ Class ______________ Date ______________

Note-taking Workbook

Biology and You

Section: What Is Biology?

THE STUDY OF LIFE

Key Idea: Biology includes ______________________________

____

Biology is ______________________________

____

Additional notes about The Study of Life: ______________________________

____

____

PROPERTIES OF LIFE

Key Idea: The seven properties of life are ______________________________

____

A **cell** is ______________________________

____

Homeostasis is ______________________________

____

Metabolism is ______________________________

____

Reproduction is the process of ______________________________

Heredity is ______________________________

____

Evolution is ______________________________

____

Reading Check: How is heredity related to evolution? ____________

____

Additional notes about the Properties of Life: ____________

____

____

Name ______________________ Class ______________ Date ______________

Note-taking Workbook

Applications of Biology

Section: Health in the 21st Century

MEETING THE CHALLENGE

Key Idea: Biologists combine ______________ and ______________ from many different fields of science to help ______________
____.

Reading Check: Why is the cholera bacterium a pathogen? ______________
____
____

Additional notes about Solving the Riddle: ______________
____
____

Additional notes about Using Models: ______________
____
____

Additional notes about A Low-Tech Solution: ______________
____
____

Reading Check: Relate water temperature to cholera outbreaks. ______________
____
____

DISEASE IN A CHANGING WORLD

Key Idea: As scientists learn more about the nature of ______________ our ability to prevent and treat ______________________.

Epidemiology is the study of ______________________________

____

Vaccination is ______________________________

____

Genetics is the science of ______________________________

____

A **genome** is ______________________________

____

Additional notes about Disease "Conquered" by Biology: ____________

____

____

Additional notes about New Insights into Disease: ____________

____

____

Reading Check: How do vaccination programs prevent diseases? _____

____

____

BIOLOGY AND HUMAN POTENTIAL

Key Idea: As our understanding of ______________________________

increases, humans will ______________________________.

The word **device** means ______________________________

____

Additional notes about Assistive Technologies: ____________

____

____

Additional notes about Battlefield Medicine: ____________

____

____

Reading Check: Give an example of an assistive technology. ________

____

____

Name ______________________ Class ______________ Date ______________

Note-taking Workbook

Applications of Biology

Section: Biology, Technology and Society

BIOTECHNOLOGY AROUND US

Key Idea: In agriculture, genetic engineering is used to ______________________________

____.

Genetic engineering is ____

____

Additional notes about Biotechnology Around Us: ______________________

____

____

APPLICATIONS OF BIOLOGICAL RESEARCH

Key Idea: Tools such as ______________________________________

have expanded the potential applications of ______________________________.

The word **process** means ______________________________________.

Additional notes about Biotechnology and Scientific Research: __________

____

____

Additional notes about Nanotechnology: ______________________________

____

____

Additional notes about Biomolecular Materials: ______________________

____

____

Additional notes about Biomimetics: ______________________________

____

____

Additional notes about Adapting Tools and Methods: ________________

____

____

Reading Check: How is the lobster-eye telescope unique from other types of telescopes? ________________________________

____

BIOLOGY, FORENSICS, AND PUBLIC SAFETY

Key idea: Because biological factors, such as ________________ ________________ are unique, they can be used to ________________ ____.

Biometrics is ________________________________

____

Additional notes about The Types of Fingerprinting: ________________

____

____

Additional notes about Other Forms of Biometric Identification: ________

____

____

Additional notes about Preventing Bioterrorism: ________________

____

____

Reading Check: What is DNA fingerprinting, and how is it used to identify someone?________________________________

____

THE ETHICS OF BIOTECHNOLOGY

Key idea: Advances in ________________ raise ethical concerns that must be addressed, both by ________________ and by ________________.

Additional notes about Manipulating DNA: ________________

____

____

Additional notes about Personal Security: ________________

____

____

Reading Check: What are some ethical concerns faced by society that relate to genetically modified organisms? ________________

____

Name ______________________________ Class __________________ Date ________________

Note-taking Workbook

Applications of Biology

Section: Biology and Environment

A LOST WORLD

Key idea: Biological research helps ______________________________

________________ the environment. We learn how to ________________

the environment by learning more about what affects it.

Ecology is: ______________________________

____

Environmental science is: ______________________________

____

Additional notes about A Lost World: ______________________________

____

____

TECHNOLOGY IN ENVIRONMENTAL SCIENCE

Key idea: Tools such as ________________, ____________________ ,

and ________________ are used to study and protect the environment.

Additional notes about Satellite Tagging: ______________________________

____

____

Additional notes about Geographic Information Systems: ____________

____

____

Additional notes about Genetic Tools: ______________________________

____

____

Reading Check: How does a GIS allow scientists to work together?

____

____

CITIZEN SCIENTISTS

Key idea: Biologists rely on the contributions of ______________

to help develop solutions for ______________.

The word **contribution** means ______________

____

Additional notes about Environmental Clubs: ______________

____

____

Additional notes about Getting Involved: ______________

____

____

Reading Check: What is the mission of the SWCC? ______________

____

Name ______________________ Class ______________ Date ______________

Note-taking Workbook

Chemistry of Life

Section: Matter and Substances

ATOMS

Key Idea: All matter is made up of ________________. An ____________ has a positively charged ________________ surrounded by ____________ ____.

An **atom** is ____
____

An **element** is ____
____

Additional notes about Atomic Structure: ____________________
____
____
____

Additional notes about Elements: ______________________________
____
____
____

Reading Check: What is a proton? ____________________________
____

CHEMICAL BONDS

Key Idea: Chemical bonds form between groups of ________________ because ________________ become stable when they have ____________ ________________ in a valence shell.

Additional notes about Covalent Bonding: ______________________

____

____

Additional notes about Ionic Bonding: ______________________

____

____

Reading Check: What is a chemical bond? ______________________

____

POLARITY

Key Idea: Hydrogen bonding plays an important role in ______________

____.

Additional notes about Solubility: ______________________

____

____

Additional notes about Hydrogen Bonds: ______________________

____

____

Reading Check: Why does salt dissolve in water? ______________________

____

Make a comparison table that compares covalent bonding and ionic bonding.

	Covalent bonding	Ionic bonding

Name ______________________ Class ______________ Date ______________

Note-taking Workbook

Chemistry of Life

Section: Water and Solutions

PROPERTIES OF WATER

Key Idea: Most of the unique properties of ______________ result because ______________________________________.

Cohesion is ______________________________________

____

Adhesion is ______________________________________

____

Additional notes about Properties of Water: ______________

____

____

SOLUTIONS

Key Idea: In solutions, some substances change the balance of __________.

A **solution** is a ______________________________________

____

An **acid** is ______________________________________

____

A **base** is ______________________________________

____

pH is ______________________________________

____

A **buffer** is ______________________________________

____

Additional notes about Acids and Bases: ______________________________

____

____

Additional notes about pH and Buffers: ______________________________

____

____

Write a sentence that uses the everyday meaning of the word *basic.* ________

____

Write a sentence that uses the scientific meaning of the word *basic.* ________

____

Name ______________________ Class ______________ Date ______________

Note-taking Workbook

Chemistry of Life

Section: Carbon Compounds

BUILDING BLOCKS OF CELLS

Key Idea: Large, complex ______________ are built from a few smaller, simpler, repeating units arranged in a precise way.

Additional notes about Carbon Compounds: ______________

____

____

Reading Check: What element is the basis of biomolecules? __________

____

CARBOHYDRATES

Key Idea: Cells use carbohydrates for sources of ______________

____.

A **carbohydrate** is a ______________________________

____

A **lipid** is a ______________________________

____

Additional notes about Energy Supply: ______________

____

____

Additional notes about Structural Support: ______________

____

____

Additional notes about Cell Recognition: ______________________

____

____

Reading Check: What is the basic unit of a carbohydrate? __________

____

The word **vary** means ______________________________.

LIPIDS

Key Idea: The main function of lipids include ______________________

____.

Additional notes about Energy Stores: ______________________

____

____

Additional notes about Water Barriers: ______________________

____

____

PROTEINS

Key idea: Proteins are chains of ______________ that twist and fold into certain shapes that determine ______________________________.

A **protein** is an ______________________________________

____

An **amino acid** is a ______________________________________

____

Additional notes about Amino Acids: ______________________

____

____

Additional notes about Levels of Structure: ______________________

____

____

Reading Check: What is a protein's primary structure? ______________

____

Describe the levels of protein structure, and identify the quantifier used to describe each level. ______________________

____

NUCLEIC ACIDS

Key idea: Nucleic acids ______________ and ______________

______________________.

A **nucleic acid** is ______________________

____

A **nucleotide** is an organic compound that ______________

____

DNA is ______________________

____

RNA is ______________________

____

ATP is ______________________

____

Additional notes about Hereditary Information: ______________

____

____

Additional notes about Energy Carriers: ______________

____

Name ______________________ Class ______________ Date ______________

Note-taking Workbook

Chemistry of Life

Section: Energy and Metabolism

CHANGING MATTER

Key Idea: Living things use different ______________ to get the energy needed for ______________________________.

Additional notes about Conservation of Mass: ______________________________

____

____

Additional notes about Conservation of Energy: ______________________________

____

____

CHEMICAL REACTIONS

Key idea: Chemical reactions can occur only when ______________________________

______________________________.

Energy is ______________________________

____

A **reactant** is a ______________________________

____

A **product** is a ______________________________

____

Activation energy is ______________________________

____

Additional notes about Activation Energy: ______________________________

____

____

Additional notes about Alignment: ______________________________

____

____

Reading Check: What causes particles to repel other particles? ______

____

BIOLOGICAL REACTIONS

Key Idea: By assisting in necessary biochemical reactions, ________ help organisms maintain ______________________.

An **enzyme** is ______________________________

____

An **active state** is ______________________________

____

Substrate is ______________________________

____

Additional notes about Enzymes: ______________________________

____

____

Additional notes about Metabolism: ______________________________

____

____

Reading Check: Why is the shape of an enzyme important? ________

____

____

Name ______________________ Class ______________ Date ______________

Note-taking Workbook

Ecosystems

Section: What is an Ecosystem?

ECOSYSTEMS

Key Idea: An ecosystem includes a community of ______________ and their ______________________________.

A **community** is ______________________________

____

An **ecosystem** is ______________________________

____

A **habitat** is ______________________________

____

Biodiversity is ______________________________

____

Additional notes about Community of Organisms: ______________

____

____

Additional notes about Physical Factors: ______________

____

____

Additional notes about Biodiversity: ______________

____

____

Reading Check: List three examples of physical parts of an ecosystem.

____

____

SUCCESSION

Key Idea: An ecosystem responds to ______________ in a way that the ecosystem is ______________________________.

Succession is ______________________________

____

Additional notes about Change in an Ecosystem: ______________

____

____

Additional notes about Equilibrium: ______________

____

____

Reading check: Why are pioneer species helpful to other species? ______

____

____

MAJOR BIOLOGICAL COMMUNITIES

Key idea: Two factors of ______________ that determine biomes are ______________________________.

Climate is ______________________________

____

A **biome** is a ______________________________

____

The word **range** means ______________________________

____

Additional notes about Major Biological Communities: ______________

____

____

TERRESTRIAL BIOMES

Key Idea: Earth's major terrestrial biomes can be grouped by __________ into ______________________________.

Additional notes about Tropical Biomes: ______________________________

____

____

Additional notes about Temperate Biomes: ______________________________

____

____

Additional notes about High Latitude Biomes: ______________________________

____

____

Reading Check: In what latitudes are savannas found? __________

____

____

AQUATIC ECOSYSTEMS

Key Idea: Aquatic ecosystems are organized into ______________, ______________, ______________, and ______________.

Additional notes about Aquatic Ecosystems: ______________________________

____

____

Use a dictionary to find the meanings of the word *aquatic* and *ecosystem*. Use the definitions to write your own definition of *aquatic ecosystems*. __________

____

Reading Check: Which aquatic ecosystems have salt water? __________

____

Name ______________________ Class ____________ Date ____________

Note-taking Workbook

Ecosystems

Section: Energy Flow in Ecosystems

TROPHIC LEVELS

Key Idea: In an ecosystem, energy flows from the ______________ to ______________ to ______________ to ______________ .

A **producer** is ______________________________

A **consumer** is ______________________________

A **decomposer** is ______________________________

A **trophic level** is ______________________________

Additional notes about Food Chains: ______________________________

Additional notes about Food Web: ______________________________

Reading check: Where do consumers get their energy? ______________

LOSS OF ENERGY

Key Idea: Energy is stored at each link in the ______________. But some energy that is used dissipates as ______________ into the environment and is not ______________.

An **energy pyramid** is ____

____

Additional notes about The Ten Percent Rule: ______________

____

____

Additional notes about Energy Pyramid: ____________________

____

____

Reading Check: When energy is transferred from one trophic level to another, where does 90% of the energy go? ____________________

____

If the prairie dog level (second level) in a food pyramid contains 35,000 units of energy, how much of that energy can be stored in the eagle level (third level) of the food pyramid? ____

Name ______________________ Class ______________ Date ______________

Note-Taking Workbook

Ecosystems

Section: Cycling of Matter

WATER CYCLE

Key idea: The water cycle continuously moves water between the __________, the ________________, and the ____________________.

Additional notes about Water Cycle: ______________________________

____

____

CARBON AND OXYGEN CYLCES

Key idea: ________________, ________________, and __________ play an important role in ______________________________ through an ecosystem.

Carbon cycle is ____

____

Respiration is ____

____

Additional notes about Carbon and Oxygen Cycles: ________________

____

____

Reading Check: How does respiration play a role in cycling carbon and oxygen through an ecosystem? ______________________________

____

Explain how the carbon cycle and the oxygen cycle are similar. __________

____

Explain how they are different. ______________________________

____

NITROGEN CYCLE

Key idea: Nitrogen must be cycled through an ________________ so that nitrogen is available for organisms to ________________.

The **nitrogen cycle** is ____

____

The word **convert** means ____

____

Additional notes about Nitrogen Cycle: ________________________

____

____

Reading Check: Explain the role of bacteria in the nitrogen cycle.

____

____

PHOSPHORUS CYCLE

Key idea: Like ________________, ________________,

________________, and ________________, phosphorus must be cycled in order for an ecosystem to support life.

The **phosphorus cycle** is ____

____

Additional notes about Phosphorus Cycle: ________________________

____

____

Reading Check: How is phosphorus passed from soil to plants? ________

____

____

Name ____________________ Class ______________ Date ______________

Note-taking Workbook

Populations and Communities

Section: Populations

WHAT IS A POPULATION?

Key Idea: Understanding ________________ is important because populations of different species ________________ one another, including ________________ populations.

A **population** is ____

____

Additional notes about What Is a Population?: ________________

____

____

Reading Check: What distinguishes one zebra population from another zebra population? ____

____

POPULATION GROWTH

Key idea: Two major models of population growth are ________________ growth and ________________ growth.

Carrying capacity is ____

____

Additional notes about Exponential Growth: ________________

____

____

Additional notes about Logistic Growth: ________________

____

____

Reading Check: What are the characteristics of a population that grows exponentially? ____________________

____

FACTORS THAT AFFECT POPULATION SIZE

Key Idea: Water, food, predators and human activity are a few of many factors that affect ____________________.

The word **affect** means ____________________

____

Additional notes about Factors That Affect Population Size: ____________

____

____

Write down the definition of the words *biotic* and *abiotic.* ____________

____

Then, write down what you think that bio- means. ____________

Use a dictionary to check your answer.

Reading Check: Describe the differences between biotic and abiotic factors. ____________________

____

HUMAN POPULATION

Key idea: Better ____________________,

____________, and ____________ are a few ways that science and technology have decreased the death rate of the human population.

Additional notes about Historic Growth: ____________________

____

____

Additional notes about Science and Technology: ______________________

____

____

Reading Check: How have advances in technology allowed the human population to grow faster? ______________________

____

Name ________________ Class ____________ Date ____________

Note-taking Workbook

Populations and Communities

Section: Interactions in Communities

PREDATOR-PREY INSTINCTS

Key Idea: Species that involve ______________________ or ______________________ relationships often develop ____________ in response to one another.

Predation is ____

____

Coevolution is ____

____

Parasitism is ____

____

Additional notes about Interactions in Communities: ____________

____

____

Reading Check: Identify one way in which herbivores and plants coevolve.

____

____

OTHER INTERACTIONS

Key Idea: Mutualism and commensalisms are two kinds of ____________ ____________ in which at least one species benefits.

Symbiosis is ____

____

Mutualism is ____

____

Commensalism is ____

____

Additional notes about Mutualism: ____________________________

____

____

Additional notes about Commensalism: _________________________

____

____

Reading Check: Compare mutualism and commensalisms. ________

____

____

Make a Venn diagram to help you compare the similarities and differences between predators, parasites and herbivores.

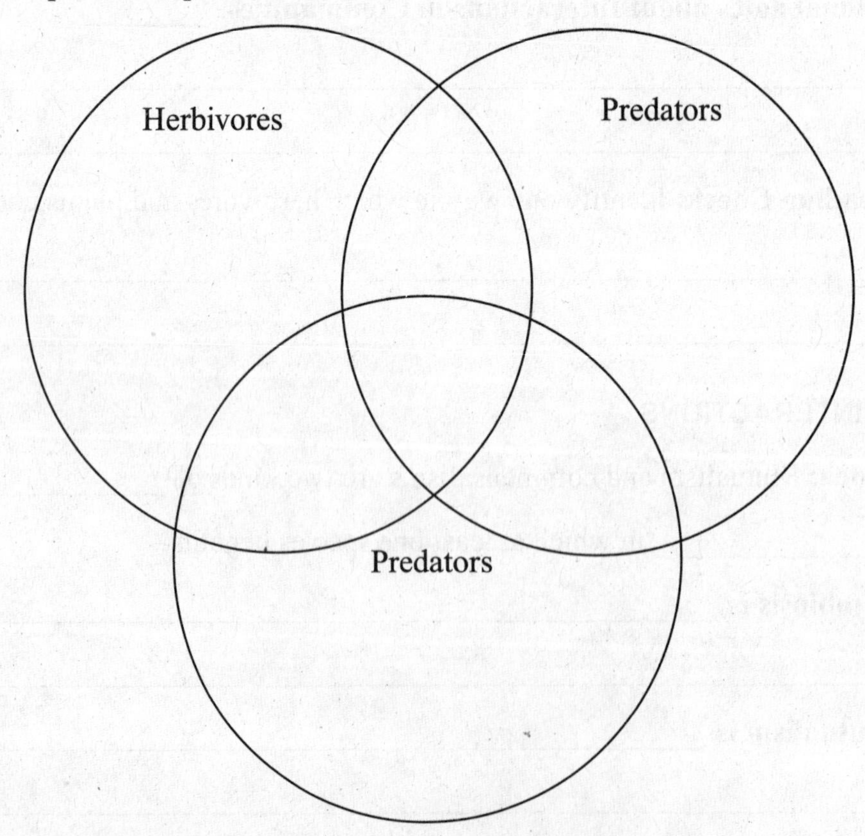

Name ______________________________ Class ________________ Date ________________

Note-Taking Workbook

Populations and Communities

Section: Shaping Communities

CARVING A NICHE

Key idea: A niche includes the role that the ________________ plays in the community.

Additional notes about Carving a Niche: ________________

____

____

Reading Check: How is a niche different from that of a habitat? ________

____

____

COMPETING FOR RESOURCES

Key idea: Competition for resources between ________________ shapes a species' fundamental ________________.

A **niche** is the ______________________________

____

A **fundamental niche** is ______________________________

____

A **realized niche** is ______________________________

____

Competitive exclusion is ______________________________

____

The word **potential** means ______________________________

____

Additional notes about Competitive Exclusion: ______________________

____

____

Additional notes about Dividing Resources: ______________________

____

____

Reading Check: How might two different species divide resources?

____

____

ECOSYSTEM RESILIENCY

Key Idea: Interactions between ________________ and the number of ________________ in an ecosystem add to the resiliency of an ecosystem.

A **keystone species** is ______________________

____

Additional notes about Ecosystem Resiliency: ______________________

____

____

Using the term *keystone species,* write a sentence with a prediction based on a condition. ______________________

____

Reading Check: List two factors that contribute to the resiliency of an ecosystem. ______________________

____

Name ______________________ Class ______________ Date ____________

Note-taking Workbook

The Environment

Section: An Interconnected Plant

HUMANS AND THE ENVIRONMENT

Key Idea: Humans are a part of the environment and can affect the

________________ of the environment.

Additional notes about Humans and the Environment: ____________

____

____

____

Reading Check: How is Earth an interconnected planet? __________

____

RESOURCES

Key Idea: Earth's resources are described as ________________ and

____.

A **fossil fuel** is ____

____

Additional notes about Renewable Resources: ________________

____

____

Additional notes about Nonrenewable Resources: ______________

____

____

Reading Check: Explain why natural gas is a nonrenewable resource.

____

____

THE ENVIRONMENT AND HEALTH

Key Idea: Pollution and habitat destruction destroy the resources needed to live, such as ______________________________.

Additional notes about The Environment and Health: ______________________________

____

____

Look up the suffix *–tion* in the dictionary. Also, look up the words *pollute* and *destroy*, then write your own definition for *pollution* and *destruction*. __________

____

____

Name ______________________ Class ______________ Date ______________

Note-taking Workbook

The Environment

Section: Environmental Issues

AIR POLLUTION

Key Idea: Air pollution causes ______________________________

____.

Acid rain is ______________________________

____

Global warming is ______________________________

____

Additional notes about Air Pollution: ______________________

____

____

GLOBAL WARMING

Key Idea: Burning fossil fuels increases the amount of ______________ in the atmosphere, which may be responsible for ______________________.

The **greenhouse effect** is ______________________________

____

Additional notes about Effects of Global Warming: ______________

____

____

Reading Check: How might the burning of fossil fuels affect climate?

____

____

WATER POLLUTION

Key Idea: Water pollution can come from ______________________

____

____.

Additional notes about Water Pollution: ______________________

____

____

Reading Check: List three sources of water pollution. ____________

____

SOIL DAMAGE

Key Idea: Without ______________________________, we will be unable to grow crops to feed ourselves or the livestock we depend on.

Erosion is ____

____

Additional notes about Soil Damage: ______________________

____

____

A lake in your state has had hundreds of dead fish wash up on shore. Write your own hypothesis that might explain why so many fish in the lake died. ________

____

____

Reading Check: How does erosion damage soil? ______________

____

ECOSYSTEM DISRUPTION

Key Idea: Ecosystem disruptions can result in ______________________________

____

____.

Deforestation is the ____

____

Biodiversity is ____

____

Extinction is ____

____

The word **sustain** means ______________________________________

____

Additional notes about Habitat Destruction: __________________________

____

____

Additional notes about Loss of Biodiversity: __________________________

____

____

Additional notes about Invasive Species: ______________________________

____

____

Additional notes about Extinction: ____________________________________

____

____

Reading Check: How has the introduction of the zebra mussel into the Great Lakes affected humans? ______________________________

____

Name ______________________ Class ______________ Date ______________

The Environment

Section: Environmental Solutions

CONSERVATION AND RESTORATION

Key Idea: Conservation involves ______________________.

Restoration involves ______________________.

Additional notes about Conservation and Restoration: ______________________

____

____

Reading Check: What is the difference between restoration and conservation? ______________________

____

REDUCE RESOURCE USE

Key Idea: We can reduce our use of resources, such as ______________ and ______________________ for energy. We can reuse goods rather than ______________ of them and ______________ waste to help protect the environment.

Recycling is ______________________

____

The word **impact** means ______________________

____

Additional notes about Reduce Resource Use: ______________________

____

____

Reading Check: What are three ways that you can reduce your use of resources? ______________________

____

TECHNOLOGY

Key Idea: Research and technology can help protect our environment by providing ____.

Additional notes about Technology: ______________________________

____

____

Reading Check: How can fuel-efficient hybrid cars help solve environmental problems? ______________________________________

ENVIRONMENTAL AWARENESS

Key Idea: Education makes people more ______________________________ and shows people how they can ______________________________.

Expressing support, or __________________, for efforts to protect the environment can help ______________________________________.

Additional notes about Environmental Awareness: ______________________

____

____

Reading Check: How can advocacy and education help solve environmental problems? ______________________________________

Make a Venn diagram to help you compare the similarities and differences between advocacy and education relating to environmental science.

Advocacy

Education

PLANNING FOR THE FUTURE

Key Idea: Careful planning for the future can help us avoid ______________

______________ and can help us ______________________

____.

Additional notes about Planning for the Future: ______________

____

____

Reading Check: Why do we need to evaluate effects of development before following through with the development? ______________

____

Name ______________________ Class ______________ Date ____________

Note-taking Workbook

Cell Structure

Section: Introduction to Cells

THE DISCOVERY OF CELLS

Key Idea: Microscope observations of organisms led to the discovery of the basic characteristics common to ______________________.

Additional notes about Cell Theory: ______________________

____

____

Reading Check: How powerful was Hooke's microscope? ______________________

____

LOOKING AT CELLS

Key Idea: A cell's shape reflects the ______________________.

The word **dimension** means ______________________

____

Additional notes about Cell Size: ______________________

____

____

Additional notes about Cell Shape: ______________________

____

____

Reading Check: How does a cell's size affect the cell's function? ______________________

____

CELL FEATURES

Key Idea: All cells have ______________________

______________________.

A **cell membrane** is ____

____

Cytoplasm is ____

____

A **ribosome** is ____

____

A **prokaryote** is ____

____

A **eukaryote** is an ____

____

The **nucleus** is ____

____

An **organelle** is ____

____

Additional notes about Features of Prokaryotic Cells: ______________

____

____

Additional notes about Features of Eukaryotic Cells: ______________

____

____

Reading Check: What is a ribosome? ______________________

____

The root *kary* means "kernel," which describes the nucleus. *Eu-* means "true," so a eukaryotic cell has a true nucleus. If *pro-* means "before," what does *prokaryotic* mean? ________________________________

____

Name ______________________________ Class ________________ Date ________________

Note-taking Workbook

Cell Structure

Section: Inside the Eukaryotic Cell

THE FRAMEWORK OF THE CELL

Key Idea: The ____________________ helps the cell move, keep its shape, and organize its parts.

Additional notes about The Framework of the Cell: ________________

____

____

DIRECTING CELLULAR ACTIVITY

Key Idea: DNA instructions are copied as ____________________, which leave the nucleus. In the cytoplasm, ____________________ use the ____________________ to assemble ____________________.

The word **assemble** means ____

____

Additional notes about Nucleus: ____________________________________

____

____

Additional notes about Ribosomes: __________________________________

____

____

Reading Check: What kind of protein do "free" ribosomes make?

____

____

PROTEIN PROCESSING

Key Idea: The ______________________ and the ______________________ are organelles that prepare proteins for extracellular export.

A **vesicle** is a ______________________

____

The **endoplasmic reticulum** is a ______________________

____

The **Golgi apparatus** is ______________________

____

Additional notes about Endoplasmic Reticulum: ______________________

____

____

Additional notes about Golgi Apparatus: ______________________

____

____

Make a process chart that shows how the cell digests food particles.

Step 1:	→	Step 2:	→	Step 3:

STORAGE AND MAINTENANCE

Key Idea: ______________ help maintain homeostasis by storing and releasing various substances as the cell needs them.

A **vacuole** is a ______________________

____

Additional notes about Lysosome: ________________________________

____

____

Additional notes about Central Vacuole: ___________________________

____

____

Additional notes about Other Vacuole: _____________________________

____

____

ENERGY PRODUCTION

Key Idea: The energy for cellular functions is produced by chemical reactions that occur in the ________________ and ________________.

The **chloroplast** is ____

____

The **mitochondrion** is ____

____

Additional notes about Chloroplasts: _______________________________

____

____

Additional notes about Mitochondria: _______________________________

____

____

Reading Check: In what kinds of cells are mitochondria found? ________

____

Name ______________________ Class ______________ Date ______________

Note-taking Workbook

Cell Structure

Section: From Cell to Organism

DIVERSITY IN CELLS

Key Idea: The different ______________ and ______________ of cells enable organisms to function in unique ways in different environments.

The **flagellum** is ______________________________

____

Additional notes about Diversity in Prokaryotes: ______________

____

____

Additional notes about Diversity in Eukaryotic Cells: ______________

____

____

Reading Check: What are flagella? ______________________________

____

LEVELS OF ORGANIZATION

Key Idea: Plants and animals have many highly specialized cells that are arranged into ______________________________.

Tissue is ______________________________

____

An **organ** is ______________________________

____

An **organ system** is ______________________________

____

Additional notes about Tissues: ______________

____

____

Additional notes about Organs: ______________

____

____

Additional notes about Organ System: ______________

____

____

Write a simile comparing each level of organization to a part of your textbook.

(Hint: Cell are like letters.) ______________

____

____

BODY TYPES

Key Idea: A ______________ is composed of many individual, permanently associated cells that coordinate their activities.

Colonial organism is ______________

____

Additional notes about Cell Groups: ______________

____

____

Additional notes about Multicellularity: ______________

____

____

Reading Check: What is differentiation? ______________

____

Name ______________________ Class ______________ Date ______________

Note-taking Workbook

Cells and Their Environment

Section: Cell Membrane

HOMEOSTASIS

Key Idea: One way that a cell maintains homeostasis is by ______________

____.

Additional notes about Homeostasis: ______________________

____

____

Reading Check: What are some roles of the cell membrane? __________

____

LIPID BILAYER

Key Idea: The phospholipids form a barrier through which only __________

____ can pass.

A **phospholipids** is a ______________________________

____

The **lipid bilayer** is ______________________________

____

Additional notes about Structure: ______________________

____

____

Additional notes about Barrier: ______________________

____

____

MEMBRANE PROTEINS

Key Idea: Proteins in the cell membrane include ______________________________

____.

Cell-Surface Markers ____

____

Receptor Proteins ____

____

Enzymes ____

____

Transport Proteins ____

____

Additional notes about Proteins in Lipids: ______________________

____

____

Additional notes about Types of Proteins: ______________________

____

____

Reading Check: Why can't ions pass through the lipid bilayer? __________

____

Make a four-corner fold to compare four types of proteins found in the cell membrane.

Cells and Their Environment

Section: Cell Transport

PASSIVE TRANSPORT

Key Idea: In passive transport, substances cross the cell membrane down their ______________________.

Equilibrium is ______________________

____

The **concentration gradient** is ______________________

____

Diffusion is ______________________

____

The **carrier protein** is ______________________

____

Additional notes about Simple Diffusion: ______________________

____

____

Additional notes about Facilitated Diffusion: ______________________

____

____

Reading Check: Why does oxygen diffuse into the cell? ______________________

____

OSMOSIS

Key Idea: Osmosis allows cells to ______________________ as their environment changes.

Osmosis is ____

____

Additional notes about Water Channels: ____

____

Additional notes about Predicting Water Movement: ____

____

Additional notes about Effects of Osmosis: ____

____

The prefix *hyper-* means "higher than," and *hypertonic* means "higher concentration." If *hypo-* means "lower than," what does *hypotonic* mean? ____

____

ACTIVE TRANSPORT

Key Idea: Active transport requires ______________ to move ______________ against their ______________.

The **sodium-potassium pump** is a ____

____

The word **release** means ____

____

Additional notes about Pumps: ____

____

____

Additional notes about Vesicles: ____

____

____

Reading Check: What is the structure of the vesicle membrane? ________

____

Name ______________________ Class ______________ Date ______________

Note-taking Workbook

Cells and Their Environment

Section: Cell Communication

SENDING SIGNALS

Key Idea: Cells communicate and coordinate activity by ______________ ______________ that carry information to other cells.

A **signal** is ______________________________
____

Additional notes about Targets: ______________________________
____
____

Additional notes about Environmental Signals: ______________________________
____
____

Reading Check: Compare the targets of signaling hormones and nerve cells. ______________________________
____

RECEIVING SIGNALS

Key Idea: A ______________________________ binds only to signals that match the specific shape of its binding site.

A **receptor protein** is ______________________________
____

Additional notes about Binding Specificity: ______________________________
____
____

Additional notes about Effect: ______________________________
____

RESPONDING TO SIGNALS

Key Idea: The cell may respond to a signal by ______________________ ______________________ by activating enzymes, or by forming a second messenger.

A **second messenger** is a ____
____

Additional notes about Responding to Signals: ______________________
____
____

Reading Check: How does membrane permeability change? ____________
____
____

Name ______________________ Class ______________ Date ______________

Note-taking Workbook

Photosynthesis and Cellular Respiration

Section: Energy in Living Systems

CHEMICAL ENERGY

Key Idea: Organisms use and store energy in the ______________________ of organic compounds. Almost all of the energy in ________________ comes from ________________.

Photosynthesis is the process by which ______________________________

____

Additional notes about Chemical Energy: ______________________________

____

____

Reading Check: Why do organisms need a constant supply of energy?

____

____

METABOLISM AND THE CARBON CYCLE

Key Idea: Metabolism involves either using energy to ________________ organic molecules or ______________________________ organic molecules in which energy is store.

Cellular respiration is the ______________________________

____

ATP is ____

____

Additional notes about Photosynthesis: ______________________________

____

____

Additional notes about Cellular Respiration: ______________________

____

____

Reading Check: How is solar energy related to the carbon cycle? ________

____

TRANSFERRING ENERGY

Key Idea: In cells, chemical energy is gradually released in a series of

______________________________ that are assisted by

______________.

ATP synthase is ____

____

The **electron transport chain** is ____________________________________

____

The word **process** means ____

____

Additional notes about ATP: ____

____

____

Additional notes about ATP Synthase: ______________________________

____

____

Additional notes about Hydrogen Ion Pumps: ________________________

____

Reading Check: How is ATP used inside a cell? ______________________

____

Name ______________________ Class ______________ Date ______________

Note-taking Workbook

Photosynthesis and Cellular Respiration

Section: Photosynthesis

HARVESTING LIGHT ENERGY

Key Idea: In plants, light energy is harvested by ______________ that are located in the ______________________________ of ______________.

A **thylakoid** is ____

____

A **pigment** is ____

____

Chlorophyll is ____

____

Additional notes about Electromagnetic Radiation: ______________

____

____

Additional notes about Pigments: ______________________________

____

____

Additional notes about Electron Carriers: ______________________

____

____

Reading Check: Describe the structure of a chloroplast. ______________

____

TWO ELECTRON TRANSPORT CHAINS

Key Idea: During photosynthesis, one electron transport chain provides energy to make ______________________, while the other provides energy to make ______________________.

Steps in the Electron Transport Chains

Step 1: ____

Step 2: ____

Step 3: ____

Step 4: ____

Step 5: ____

Additional notes about Producing ATP: ____

____

____

Additional notes about Producing NADPH: ____

____

____

Reading Check: Summarize how ATP and NADPH are formed during photosynthesis. ____

____

Use spatial language to describe production of ATP and NADPH during photosynthesis. ____

____

____

PRODUCING SUGAR

Key Idea: In the final stage of photosynthesis, ATP and NADPH are used to produce energy-storing ______________________________ from the carbon in carbon dioxide.

The **Calvin Cycle** is ______________________________
____

The word **method** means ______________________________
____

Additional notes about Producing Sugar: ______________________________
____
____

FACTORS THAT AFFECT PHOTOSYNTHESIS

Key Idea: ______________________________, ______________
______________, and ______________ are three environmental factors that affect photosynthesis.

Additional notes about Factors that Affect Photosynthesis: ______________
____
____

Reading Check: How does temperature affect photosynthesis? ______________
____
____

Name ______________________ Class ______________ Date ______________

Note-taking Workbook

Photosynthesis and Cellular Respiration

Section: Cellular Respiration

GLYCOLYSIS

Key Idea: The breaking of a sugar molecule by ______________ results in a net gain of ______________________.

Glycolysis is ____

____

Anaerobic describes ______________________________________

Aerobic describes ____

Steps of Glycolysis

Step 1: ____

Step 2: ____

Step 3: ____

Additional notes about Steps of Glycolysis: ______________________

____

____

Reading Check: What are the three products of glycolysis? __________

____

AEROBIC RESPIRATION

Key Idea: The total yield of energy-storing products from one time through the Krebs cycle is ______________, ______________, and ______________. Electron carriers transfer ______________ through the electron transport chain, which ultimately powers ______________.

The **Krebs cycle** is ____

____

Additional notes about Krebs Cycle: ______________________________

____

____

Additional notes about Electron Transport Chain: ______________

____

____

Reading Check: Why is glycolysis important to the Krebs cycle? _______

____

Make a pattern puzzle to help you remember the steps in aerobic respiration.

Step 1:	→	Step 2:	→	Step 3:	→	Step 4:

FERMETATION

Key Idea: Fermentation enables ______________ to continue supplying a cell with ______________ in ______________________.

Fermentation is ____

____

The word **transfer** means ______________________________________

____

Additional notes about Fermentation: ___________________________

____

____

____

Reading Check: Explain how fermentation recycles NAD+. __________

____

____

Name Angela Badillo Class 841 Date 11-29-13

Note-taking Workbook

Cell Growth and Division

Section: Cell Reproduction

WHY CELLS REPRODUCE

Key Idea: Because larger cells are more ________________________,

cells ________________ when they grow to a certain size.

Additional notes about Cell Size: ________________________

____

Additional notes about Cell Maintenance: ________________________

____

____

Additional notes about Making New Cells: ________________________

____

CHROMOSOMES

Key Idea: Eukaryotic ________________ is packaged into highly condensed

______________________ with the help of ______________________.

A **gene** is a molecular unit of heredity of a living organisms.

A **chromosome** is an organized structure of DNA, protein and RNA found in cells.

Chromatin is the combination of DNA and protein that make up the contents of the nucleus of a cell.

Histone is a highly alkaline proteins found in eukaryotic cell nuclei

The **nucleosome** is the basic unit of DNA packaging in eukaryotes, consisting of a segment of DNA.

A **chromatid** is one of ____

____

The **centromere** is ____

____

Additional notes about Chromosomes: ____

____

____

____

Reading Check: What is a chromatid? ____

____

The prefix *tel-* means "end." If *centromere* means "a central part," what do you think *telomere* means? ____

PREPARING FOR CELL DIVISION

Key Idea: All newly-formed cells require DNA, so before a cell divides, ______________________________ is made for each daughter cell.

The word **complex** means ____

____

Additional notes about Prokaryotes: ____

____

____

Additional notes about Eukaryotes: ____

____

____

Reading Check: Where does a prokaryotic cell begin to divide? ________

____

Name _______________ Class _______________ Date _______________

Note-taking Workbook

Cell Growth and Division

Section: Mitosis

EUKAYOTIC CELL CYCLE

Key Idea: The life of a eukaryotic cell cycles through phases of ____________, ____________, preparation for cell division, and ____________ ____________.

The **cell cycle** is ______________________________

The **interphase** is ______________________________

Mitosis is ______________________________

Cytokinesis is ______________________________

Phases of the Cell Cycle:

Phase G1: ______________________________

S: ______________________________

Phase G2: ______________________________

Mitosis: ______________________________

Cytokinesis: ______________________________

Additional notes about Eukaryotic Cell Cycle: ______________________________

Reading Check: What phases are included in interphase? ____________

STAGES OF MITOSIS

Key Idea: Mitosis is a continuous process that can be observed in four stages: ______________, ______________, ______________, and ______________.

A **spindle** is ____

A **centrosome** is an ____

Stage 1: Prophase ____

Stage 2: Metaphase ____

Stage 3: Anaphase ____

Stage 4: Telophase ____

Additional notes about Stages of Mitosis: ____

Reading Check: What is the spindle composed of? ____

CYTOKINESIS

Key Idea: During cytokinesis, the ______________ grows into the center of the cell and divides it into ____.

Each daughter cell has about half of ____.

The word **rigid** means ____

Additional notes about Cytokinesis: ____

Reading Check: What is a cell plate? ____

Name ______________________ Class ______________ Date ______________

Note-taking Workbook

Cell Growth and Division

Section: Regulation

CONTROLS

Key Idea: Cell growth and division depend on ______________________ and other ______________________.

Additional notes about Controls: ______________________

____

____

Reading Check: What are two factors that affect the cell cycle? ____________

____

CHECKPOINTS

Key Idea: Feedback signals at key ______________ in the cell cycle can ______________ or ______________ the next phase of the cell cycle.

Additional notes about Checkpoints: ______________________

____

____

____

____

Reading Check: What happens at the G2 checkpoint? ____________

____

At each checkpoint, identify a cause that would result in a delay of the next cycle. ______________________

____

____

CANCER

Key Idea: ______________________________ and ______________ can result in masses of cells that invade and destroy health tissues.

Cancer is ____

A tumor is ____

Additional notes about Loss of Control: ______________________

Additional notes about Development: ______________________

Additional notes about Treatment: ______________________

Additional notes about Prevention: ______________________

Reading Check: What causes cells to lose control of the cell cycle?

____

Name ______________________ Class ______________ Date ______________

Note-taking Workbook

Meiosis and Sexual Reproduction

Section: Reproduction

ASEXUAL REPRODUCTION

Key Idea: An individual formed by asexual reproduction is ______________ ______________ to its parent.

Additional notes about Asexual Reproduction: ______________

____

____

Reading Check: What is fragmentation? ______________

____

SEXUAL REPRODUCTION

Key Idea: In sexual reproduction, ______________________

give ______________ to produce ______________ that are genetically ______________ from their parents.

A **gamete** is ______________________________

____

A **zygote** is ______________________________

____

Additional notes about Germ Cells and Somatic Cells: ______________

____

____

Additional notes about Advantages of Sexual Reproduction: ______________

____

____

CHROMOSOME NUMBER

Key Idea: Each chromosome has ______________ of genes that play an important role in determining how an organism develops and functions.

A **diploid** is ____

____

A **haploid** is ____

____

Homologous chromosomes are ______________________________

____

Additional notes about Haploid and Diploid Cells: ______________

____

____

Additional notes about Homologous Chromosomes: ______________

____

____

Additional notes about Autosomes and Sex Chromosomes: ________

____

____

Reading Check: What kind of cells do germ cells produce? ________

Name ______________________ Class ______________ Date ______________

Note-taking Workbook

Meiosis and Sexual Reproduction

Section: Meiosis

STAGES OF MEIOSIS

Key Idea: During meiosis, a diploid cell goes through ______________ divisions to form ______________ haploid cells.

Meiosis is ____

____

Crossing-over is ____

____

Stages of Mitosis:

Stage 1: Prophase I ______________________________

Stage 2: ______________________________

Stage 3: Anaphase I ______________________________

Stage 4: ______________________________

Stage 5: ______________________________

Stage 6: Metaphase II ______________________________

Stage 7: ______________________________

Stage 8: Telophase II ______________________________

Additional notes about Meiosis I: ______________________________

____

____

Additional notes about Meiosis II: ______________________________

____

____

Reading Check: In what phase of meiosis is genetic material exchanged?

____

____

COMPARING MITOSIS AND MEIOSIS

Key Idea: Mitosis makes new cells that are used during growth, development, repair and asexual reproduction. Meiosis makes cells that enable an organism to ______________________________ and happens only in ______________________.

Additional notes about Comparing Mitosis and Meiosis: ______________________________

Reading Check: How are cells formed by mitosis different from cells formed by meiosis in relation to number of chromosomes? ______________________________

Write two sentences that compare and two sentences that contrast meiosis and mitosis. ______________________________

GENETIC VARIATION

Key Idea: Three key contributions to genetic variation are ______________________________.

The word **exist** means ______________________________

Additional notes about Crossing-Over: ______________________________

Reading Check: How can crossing-over increase genetic variation? ______________________________

Independent assortment is ______________________________

____

Additional notes about Independent Assortment: ____________

____

____

Additional notes about Random Fertilization: ______________

____

____

Name ______________________ Class ______________ Date ______________

Note-taking Workbook

Meiosis and Sexual Reproduction

Section: Multicellular Life Cycles

DIPLOID LIFE CYCLE

Key Idea: In ______________ life cycles, meiosis in germ cells of a ______________________ results in the formulation of ______________.

A **life cycle** is ____

____

Sperm are ____

____

An **ovum** is ____

____

Additional notes about Diploid Life Cycle: ______________________

____

____

Reading Check: How many gametes are formed from one female germ cell?____

____

HAPLOID LIFE CYCLE

Key Idea: In ______________ life cycles, meiosis in a diploid zygote results in ____.

Additional notes about Haploid Life Cycle: ______________________

____

____

MENDEL'S FINDINGS IN MODERN TERMS

Key Idea: Mendel's experiments showed that ________________ determines ________________.

A **genotype** is ____
____

A **phenotype** is ____
____

Homozygous describes ____
____

Heterozygous describes ____
____

Additional notes about Genotype and Phenotype: ________________
____

Additional notes about Homozygous and Heterozygous: ____________
____
____

Look up the word *phenomenon* in a dictionary. What is the meaning of the Greek root of this word? ____________________________________

How does this meaning apply to the word *phenotype* as used in biology? _____
____
____

MENDEL'S SECOND EXPERIMENTS

Key Idea: In modern terms, ______________________________
holds that during gamete formation, the alleles of each gene segregate independently.

Additional notes about Independent Assortment: ____________________
____
____

Additional notes about Genes Linked on Chromosomes: __________
____
____

Reading Check: What is a dihybrid cross? ____________________
____

Name ______________________ Class ______________ Date ______________

Note-taking Workbook

Mendel and Heredity

Section: Modeling Mendel's Laws

USING PUNNETT SQUARES

Key Idea: A ______________ shows all of the ______________ that could result from ______________.

A **Punnett square** is a ____

____

The word **contrast** means ____

____

Additional notes about Analyzing Monohybrid Crosses: ______________

____

____

Reading Check: Explain the boxes inside a Punnett square. ______________

____

Solve the following analogy:

yy : Yy :: Homozygous : ______________

USING PROBABILITY

Key Idea: Probability formulas can be used to ______________ that ______________________ will be passed on to offspring.

Probability is ____

____

The word **occur** means ____.

____

$$\text{Probability} = \frac{\text{Number of \_\_\_\_\_\_\_\_\_\_ possible outcome}}{\text{\_\_\_\_\_\_\_\_\_\_\_\_\_\_ possible outcomes}}$$

Additional notes about Calculating Probability: ______________

____

____

Reading Check: What is the probability that a heterozygous cross will produce homozygous recessive offspring? ______________________

____

USING A PEDIGREE

Key Idea: A pedigree can help answer questions about three aspects of inheritance: ____

____.

A **pedigree** is ____

____

A **genetic disorder** is ____

____

Additional notes about Using a Pedigree: ______________________

____

____

____

____

Reading Check: How can one identify a sex-linked trait? __________

____

Name ______________________ Class ______________ Date ______________

Note-taking Workbook

Mendel and Heredity

Section: Beyond Mendelian Heredity

MANY GENES, MANY ALLELES

Key Idea: The Mendelian inheritance pattern is rare in nature; other patterns include ____

____.

A **polygenic character** is ________________________________

____

Codominance is ________________________________

____

The word **various** means ________________________________

____

Additional notes about Polygenic Inheritance: ____________________

____

____

Additional notes about Incomplete Dominance: ____________________

____

____

Additional notes about Multiple Alleles: ____________________

____

____

Additional notes about Codominance: ____________________

____

____

Reading Check: How does codominance differ from incomplete dominance? ________________________________

____

GENES AFFECTED BY THE ENVIRONMENT

Key Idea: ______________ can be affected by conditions in the environment, such as ______________ and ______________.

Additional notes about Genes Affected by the Environment: ______________

GENES LINKED WITHIN CHROMOSOMES

Key Idea: During ______________, genes that are ______________ on the same chromosome are less likely to be separated than ______________.

Linked describes ______________

Additional notes about Genes Linked Within Chromosomes: ______________

Reading Check: What term describes genes that are close together on the same chromosome and that are unlikely to be separated? ______________

Name ______________________ Class ______________ Date ______________

Note-taking Workbook

DNA, RNA, and Proteins

Section: The Structure of DNA

DNA: THE GENETIC MATERIAL

Key Idea: DNA is the primary material that causes ______________,
______________________________ in related groups of organisms.

A **gene** is ____
____

DNA is ____
____

Additional notes about DNA: The Genetic Material: ______________
____

Reading Check: What are genes composed of? ______________
____

SEARCHING FOR THE GENETIC MATERIAL

Key Idea: Three major experiments led to the conclusion that __________
______________________________. These experiments were
performed by ____.

Additional notes about Griffith's Discovery of Transformation: ________
____
____

Use specific time markers to describe Griffith's experiment. ____________
____
____

Additional notes about Avery's Experiments with Nucleic Acids: ______
____
____

Steps of the Hershey-Chase Experiement: ____________________

____

____

THE SHAPE OF DNA

Key Idea: A DNA molecule is shaped like a ____________________ and is composed of two ____________________ of linked subunits.

Nucleotide is ____________________

____

Additional notes about A Winding Staircase: ____________________

____

____

Additional notes about Parts of the Nucleotide Subunits: __________

____

____

THE INFORMATION IN DNA

Key Idea: The information in DNA is contained in the order of the ______________, while the base-pairing structure allows the information to be copied.

A **purine** is a ____________________

____

A **pyrimidine** is a ____________________

____

The word **complementary** means ____________________

____

Additional notes about Nitrogenous Bases: ______________________

____

Additional notes about Base-Pairing Rules: ______________________

____

Additional notes about Complementary Sides: ______________________

____

Reading Check: How are base-pairs held together? ______________

____

DISCOVERING DNA'S STRUCTURE

Key Idea: ________________ and ________________ used information from experiments by Chargaff, Wilkins, and Franklin to determine the ______ ____.

Additional notes about Observing Patterns: Chargaff's Observations:

____

____

____

Additional notes about Using Technology: Photographs of DNA: ______

____

____

Additional notes about Watson and Crick's Model of DNA: __________

____

____

Reading Check: How was X-ray diffraction used to model the structure of DNA? ____

____

Name ______________________ Class ______________ Date ______________

Note-taking Workbook

DNA, RNA, and Proteins

Section: Replication of DNA

DNA REPLICATION

Key Idea: In DNA ______________, the DNA molecule ______________, and the two sides ________________. Then, new bases are added to each side until ______________________________________.

DNA replication is ______________________________________

____

Additional notes about DNA Replication: ______________

____

REPLICATION PROTEINS

Key Idea: During the replication of DNA, many ________________ form a machinelike complex of moving parts.

A **DNA helicase** is ______________________________________

____

A **DNA polymerase** is ____________________________________

____

Additional notes about DNA Helicase: __________________

____

____

Additional notes about DNA Polymerase: ________________

____

____

Reading Check: Why is proofreading critical during replication? ______

____

____

PROKARYOTIC AND EUKARYOTIC REPLICATION

Key Idea: In prokaryotic cells, replication starts ________________.

In eukaryotic cells, replication starts ________________

________________.

The word **distinct** means ________________

____

Additional notes about Prokaryotic DNA Replication: ________________

____

____

Additional notes about Eukaryotic DNA Replication: ________________

____

____

Reading Check: How is a “replication bubble” formed? ________________

____

In your own words, write a definition of *helicase* and *polymerase*, the names of the enzymes involved in DNA replication, based on the meanings of each term’s word parts. ________________

____

____

Name ______________________ Class ______________ Date ______________

Note-Taking Workbook

DNA, RNA, and Proteins

Section: RNA and Gene Expression

AN OVERVIEW OF GENE EXPRESSION

Key Idea: Gene expression produces proteins by ______________ and ______________. This process takes place in ______________ stages, both of which involve ______________.

RNA is ____

____

Gene expression is ____

____

Transcription is ____

____

Translation is ____

____

Additional notes about Transcription: DNA to RNA: ______________

____

____

Additional notes about Translation: RNA to Proteins: ______________

____

____

RNA: A MAJOR PLAYER

Key Idea: In cells, ______________ types of RNA complement DNA and translate the ______________ into ______________.

Additional notes about RNA Versus DNA: ______________________________

____

____

Types of RNA

Messenger RNA: ______________________

Transfer RNA: ______________________

Ribosomal RNA: ______________________

Additional Notes about Types of RNA: ______________________

____

____

Reading Check: What are the structural differences between RNA and DNA? ______________________

____

TRANSCRIPTION: READING THE GENE

Key Idea: During transcription, the information in a specific region of DNA (a gene) is transcribed, or copied, into ______________.

Additional notes about Transcription Versus Replication: ______________

____

____

Reading Check: What is the role of a promoter? ______________

____

THE GENETIC CODE: THREE-LETTER "WORDS"

Key Idea: The genetic code is based on ______________ that each represent a specific ______________.

A **codon** is a ______________________

____

Additional notes about Codons of mRNA: ______________

____

____

TRANSLATION: RNA TO PROTEINS

Key Idea: Translation occurs in a sequence of steps, involves three kinds of RNA, and results in a complete ______________.

Steps of Translation

Step 1: ______________________________

Step 2: ______________________________

Step 3: ______________________________

Step 4: ______________________________

Step 5: ______________________________

Additional notes about Translation: RNA to Proteins: ______________

____

____

Additional notes about Repeating Translation: ______________

____

____

Reading Check: How do codons and anticodons differ? ______________

____

COMPLEXITIES OF GENE EXPRESSION

Key Idea: The relationship between ______________ and their effects is complex. Despite the neatness of the genetic code, every ______________ cannot be simply linked to a single outcome.

Additional notes about Complexities of Gene Expression: ______________

____

____

Name ______________________ Class ______________ Date ______________

Note-taking Workbook

Genes in Action

Section: Mutation and Genetic Change

MUTATION: THE BASIS OF GENETIC CHANGE

Key Idea: For the most part, ______________________________

among organisms originate as some kind of ______________________.

A **mutation** is ______________________________

____

Additional notes about Causes of Mutation: ______________

____

____

Additional notes about Effects of Mutations: ______________

____

____

Reading Check: Where do new alleles comes from? ______________

____

SEVERAL KINDS OF MUTATIONS

Key Idea: Different kinds of mutations are recognized as either ______________

______________________________ or ______________________

______________________________.

Additional notes about Mutations as Changes in DNA: ______________

____

____

Additional notes about Mutations as Changes in Results of Genes:

____

____

____

Additional notes about Chromosomal Mutations: ____________________

____

____

Reading Check: Why are point mutations often silent? ____________

____

EFFECTS OF GENETIC CHANGE

Key Idea: The results of genetic change may be ____________________,

____________, or ____________; most changes are

____________ and may not be passed on to offspring.

The word **dramatic** means ____________________________

____

Additional notes about Heritable or Not: ____________________

____

____

Additional notes about Tumors and Cancers: ____________________

____

____

Additional notes about New Alleles: ____________________

____

____

Additional notes about Genetic Disorders: ____________________

____

____

Reading Check: How are mutations related to cancer? ____________

____

LARGE-SCALE GENETIC CHANGE

Key Idea: Very large-scale genetic change can occur by ______________,

______________, or ______________.

Nondisjunction is ______________________________

____

Polyploidy is ______________________________

____

Additional notes about Recombination During Crossover: ______________

____

____

____

Additional notes about Errors in Sorting Chromosomes: ______________

____

____

Reading Check: How can a child be born with extra chromosomes?

____

____

Name ______________________ Class ______________ Date ______________

Note-taking Workbook

Genes in Action

Section: Regulating Gene Expression

COMPLEXITIES OF GENE EXPRESSION

Key Idea: Cells have ________________ that regulate whether or not specific genes are expressed. Expression depends on the cell's ________________ ________________.

Additional notes about Complexities of Gene Expression: ________________

____

____

Reading Check: Are all genes expressed all the time? ____________

____

GENE REGULATION IN PROKARYOTES

Key Idea: The major form of gene regulation in prokaryotes depends upon ________________ that respond to environmental factors.

An **operon** is a ____

____

Additional notes about Interactions with the Environment: ____________

____

____

Additional notes about The *lac* Operon Example: ________________

____

____

GENE REGULATION IN EUKARYOTES

Key Idea: Gene regulation in eukaryotes in more ________________ and ________________ than gene regulation in prokaryotes.

The **transcription factor** is ______________________________

An **intron** is a ______________________________

An **exon** is ______________________________

The word **regulate** means ______________________________

Additional notes about Controlling Transcription: ______________________________

Additional notes about Processing RNA After Transcription: ______________________________

Additional notes about Processing Proteins After Translation: ______________________________

Reading Check: Which parts of gene expression can be regulated?

____

THE MANY ROLES OF PROTEINS

Key Idea: The sequence of ______________ in a protein determines its three-dimensional structure and chemical behavior.

A **domain** is ______________________________

Additional notes about Protein Structure: _______________

____

Additional notes about Proteins in Gene Expression: _______________

____

____

Reading Check: What determines a protein's shape? _______________

____

____

Name ______________________ Class ______________ Date ______________

Note-taking Workbook

Genes in Action

Section: Genome Interactions

GENOMES AND THE DIVERSITY OF LIFE

Key Idea: Comparisons among the genetic systems of many organisms reveal basic biological ______________________.

A **genome** is ______________________

____

A **plasmid** is a ______________________

____

Additional notes about Universal Code: ______________________

____

____

Additional notes about Genome Sizes: ______________________

____

____

Additional notes about DNA Versus Genes: ______________________

____

____

Additional notes about Endosymbiotic Theory: ______________________

____

____

Reading Check: What kinds of organisms have large genomes?

____

____

MOVING BEYOND CHROMOSOMES

Key Idea: Small bits of genetic material can be ________________,

________________ and ________________ by a variety of interactions.

A **transposon** is a ________________________________

____

Additional notes about Mobile Genetic Elements: ________________

____

____

Additional notes about Genetic Change: ________________________

____

____

Reading Check: How are transposons and viruses similar? __________

____

MULTICELLULAR DEVELOPMENT AND AGING

Key Idea: Each cell within a developing body will express specific genes. Gene expression depends on the cell's ________________ and

____.

Cell differentiation is ________________________________

____

Apoptosis is ________________________________

____

Additional notes about Cell Differentiation: ________________

____

____

Reading Check: What is a homeobox? ________________________

____

Additional notes about Cell Growth and Maintenance: ______________

____

____

Additional notes about Cell Death and Aging: ______________

____

____

Reading Check: What are the roles of proteins in the cell cycle?

____

____

Find out the meaning of the prefix *homeo-*. ______________

____

Write a definition for *homeotic* and *homeobox* in your own words. ______________

____

____

____

Name ______________________ Class ______________ Date ______________

Note-taking Workbook

Gene Technologies and Human Applications

Section: The Human Genome

SECRETS OF THE HUMAN GENOME

Key Idea: The sequencing of the human genome has advanced the study of

______________________ yet created new questions.

Genomics is the study of ______________________

____

Additional notes about Surprising Findings: ______________________

____

____

Reading Check: How big is the human genome? ______________________

____

APPLICATIONS OF HUMAN GENETICS

Key Idea: ______________ and ______________________

have many applications in human healthcare and society.

A **microarray** is a ______________________

____

A **DNA fingerprint** is a ______________________

____

Additional notes about Diagnosing and Preventing Disease: ______________

____

____

Reading Check: When might a person seek genetic counseling?

____

____

Additional notes about Treating Disease: ______________________

____

____

Additional notes about Identifying Individuals: ______________________

____

____

Reading Check: Why is insulin used to treat genetic diabetes?

____

____

The prefix *pharma-* means "medicine" or "drug." Use this information to analyze the meaning of the term *pharmacogenomics.* ______________________

____

____

ONGOING WORK

Key Idea: Many important questions about the ______________________ remain to be investigated and decided.

The word **implication** means ______________________

____

Additional notes about Ongoing Work: ______________________

____

____

Reading Check: Why is asthma difficult to cure? ______________________

____

Name ______________________ Class ______________ Date ______________

Note-taking Workbook

Gene Technologies and Human Applications

Section: Gene Technologies in Our Lives

MANIPULATING GENES

Key Idea: Gene technologies are now widely applied to ______________ ______________, to ______________, and to ______________.

Genetic engineering is when you take out something from an organism DNA and put into other organism DNA

Recombinant DNA are ______________

Additional notes about Genetic Engineering: ______________

Reading Check: What is a GMO? ______________

Additional notes about Everyday Applications: ______________

Additional notes about Manipulating Cell Interactions: ______________

Reading Check: What is the Bt gene used for? ______________

Name ______________________ Class ______________ Date ______________

MANIPULATING BODIES AND DEVELOPMENT

Key Idea: ________________, and ______________________ ________________ are used in research on animal development and have potential for treating certain diseases.

A **clone** is an exact copy of something

A **stem cell** is a an undifferentiated body

Additional notes about Cloning: ______________________

Additional notes about Using Stem Cells: ______________________

Reading Check: What are the two main types of stem cells? Embryonic stem cells and adult stem cells

ETHICAL AND SOCIAL ISSUES

Key Idea: ______________________ can be raised for every use of ______________________.

The word **ethical** means ______________________

Additional notes about Ethical and Social Issues: ______________________

Reading Check: What issues does the use of genetic testing use?

Name ____________________ Class ______________ Date ______________

Note-taking Workbook

Gene Technologies and Human Applications

Section: Gene Technologies in Detail

BASIC TOOLS FOR GENETIC MANIPULATION

Key Idea: The basic tools of DNA manipulation rely on the ______________ ______________________________ and are adapted from natural processes discovered in cells.

A **restriction enzyme** is ______________________________

____

DNA polymorphisms are ______________________________

____

Electrophoresis is ______________________________

____

The word **slight** means ______________________________

____

Additional notes about Restriction Enzymes: ______________

____

____

Reading Check: Which basic genetic tools were used to make the first GMOs? ______________________________

____

Additional notes about Polymorphisms: ______________

____

Additional notes about Gel Electrophoresis: ______________

____

____

Reading Check: What property of a gel does gel electrophoresis depend upon? ____

____

Additional notes about Denaturation: ____

____

____

Additional notes about Hybridization: ____

____

____

MAJOR GENE TECHNOLOGY PROCESSES

Key Idea: The major methods for working with genes use some combination of the basic tools and mechanisms of ________________________.

The **Polymerase chain reaction** is ____

____

Blotting Processes

Southern Blot: ____

____

Fingerprints and Bar Codes: ____

____

Northern Blot: ____

____

Microarrays: ____

____

Reading Check: What does "blotting" refer to? ________________________

____

DNA sequencing is ____

____

DNA Sequencing

Step 1: ____

Step 2: ____

Step 3: ____

Reading Check: When are primers used in DNA sequencing? __________

____

Gene Recombination

Step 1: ____

Step 2: ____

Step 3: ____

Step 4: ____

Step 5: ____

Additional notes about Major Gene Technology Processes: ________

____

____

Reading Check: What is a vector? ____________________

____

EXPLORING GENOMES

Key Idea: One can explore and map a genome at many levels, including

________________, ________________, ________________

________________, or ________________________________.

Bioinformatics is ____

____

Genome mapping is ________________________________

____

A **genetic library** is a ________________________________

____

Additional notes about Managing Genomic Data: ____________

____

____

Reading Check: What are the first steps of studying genomes? ________

____

Additional notes about Mapping Methods: ____________________

____

____

Additional notes about Genome Sequence Assembly: ____________

____

____

Reading Check: What are the two kinds of genetic libraries? ________

____

____

Name ______________________ Class ______________ Date ______________

Note-taking Workbook

Evolutionary Theory

Section: Developing a Theory

A THEORY TO EXPLAIN CHANGE OVER TIME

Key Idea: Modern ________________ began when ________________ presented evidence that ________________ and offered an ____________ of how ________________.

Evolution is ____

____

Additional notes about A Theory to Explain Change Over Time: ________

____

____

Reading Check: What does evolution mean in biology? ______________

____

DARWIN'S IDEAS FROM EXPERIENCE

Key Idea: Darwin's experiences provided him with evidence of ______ ____.

Artificial selection is ____

____

The word **insight** means ____

____

Additional notes about The Voyage of the *Beagle*: ________________

____

____

Additional notes about Years of Reflection: ________________

____

____

Additional notes about Breeding and Selection: ____________________

____

____

Reading Check: When did Darwin first see evidence of evolution? _____

____

DARWIN'S IDEAS FROM OTHERS

Key Idea: Darwin was influenced by ideas from the fields of ____________,

______________, ______________, and ________________.

Additional notes about Lamarckian Inheritance: ________________

____

____

Additional notes about Population Growth: ____________________

____

____

Additional notes about Geology and an Ancient Earth: __________

____

____

Reading Check: What belief did Darwin and Lamarck share? _______

____

____

Name _______________ Class _______________ Date _______________

Note-taking Workbook

Evolutionary Theory

Section: Applying Darwin's Ideas

EVOLUTION BY NATURAL SELECTION

Key Idea: Darwin's theory predicts that over time, the number of individuals that carry _______________ traits will increase in population.

Natural selection is _______________

____

Adaptation is a _______________

____

Steps of Darwin's Theory

Step 1: _______________

Step 2: _______________

Step 3: _______________

Step 4: _______________

Additional notes about Steps of Darwin's Theory: _______________

____

____

Additional notes about Selection and Adaptation: _______________

____

____

Additional notes about Publication of the Theory: _______________

____

____

Reading Check: Is natural selection the same thing as evolution? _______________

____

____

WHAT DARWIN EXPLAINED

Key Idea: Darwin presented a unifying explanation for ______________ from multiple fields of science.

A **fossil** is ______________________________

____

Homologous describes ______________________________

____

The word **infer** means ______________________________

____

Additional notes about The Fossil Record: ______________________________

____

____

Additional notes about Biogeography: ______________________________

____

Additional notes about Developmental Biology: ______________________________

____

Reading Check: Why is the fossil record incomplete? ______________

____

Additional notes about Anatomy: ______________________________

____

____

Additional notes about Biochemistry: ______________________________

____

____

Reading Check: What explains similarities in bone structure?

____

____

EVALUATING DARWIN'S IDEA

Key Idea: Darwin's work had three major strengths: ______________

____

______________.

Additional notes about Strengths: ______________

____

____

Additional notes about Weaknesses: ______________

____

____

Reading Check: What did Darwin do before publishing his ideas?

____

____

Name ______________________ Class ______________ Date ______________

Note-taking Workbook

Evolutionary Theory

Section: Beyond Darwinian Theory

DARWIN'S THEORY UPDATED

Key Idea: Discoveries since ______________, especially in ______________ have been added to his theory to explain ______________________.

Additional notes about Remaining Questions: ______________________

____

____

STUDYING EVOLUTION AT ALL SCALES

Key Idea: Because it affects every aspect of biology, scientists can study ______________ at many scales. Generally, these scales range from ______________ to ______________.

Speciation is ______________________________

____

The word **random** means ______________________

____

Additional notes about Speciation: ______________________

____

Additional notes about Processes of Microevolution: ______________

____

____

Additional notes about Patterns of Macroevolution: ______________

____

____

Reading Check: At what scales can evolution be studied? ______________

____

Name ______________________ Class ______________ Date ______________

Note-taking Workbook

Population Genetics and Speciation

Section: Genetic Variation

POPULATION GENETICS

Key Idea: ______________ can be studied by observing changes in the numbers and types of ______________ in populations.

Population genetics is the study of ______________________________

____

Additional notes about Population Genetics: ______________________

____

____

Reading Check: What do we now know about heredity that Darwin did not know? ____

____

PHENOTYPIC VARIATION

Key Idea: Biologists study ______________________ by measuring each individual in the population and then analyzing the distribution of the measurements.

Normal distribution is ______________________________

____

Additional notes about Phenotypic Variation: ______________________

____

____

Reading Check: Why do polygenic characters vary so much?

____

____

MEASURING VARIATION AND CHANGE

Key Idea: Genetic variation and change are measured in terms of the frequency of ________________ in the gene pool of a population

Additional notes about Studying Alleles: ____________________________

____

Reading Check: What is the main measure of genetic variation? _______

____

The word *normal* in science and math is often used to describe measurements that fit within a normal distribution. What does a doctor mean when talking about "normal height" for a person of your age? ____________________

____

$$(\text{frequency of } E) + (\text{frequency of } e) = 1$$

$$\frac{(\text{count of } E)}{(\text{total})} + \frac{(\text{count of } e)}{(\text{total})} = 1$$

Additional notes about Tracking Frequencies: ____________________

____

Reading Check: What is the sum of all allele frequencies for any one gene?

____

SOURCES OF GENETIC VARIATION

Key Idea: The major source of new ________________ in natural populations is ________________ in ________________ cells.

The word **generate** means ______________________________

____

Additional notes about Sources of Genetic Variation: ______________

____

Reading Check: Why is mutation so important: ____________________

____

Name ______________________ Class ______________ Date ______________

Note-taking Workbook

Population Genetics and Speciation

Section: Genetic Change

EQUILIBRIUM AND CHANGE

Key Idea: The ______________________ predicts that the frequencies of alleles and genotypes in a population will not change unless at least ______________________ acts upon the population.

Genetic equilibrium is ______________________

____

Additional notes about Measuring Change: ______________________

____

Additional notes about Hardy-Weinberg Principle: ______________________

____

Additional notes about Forces of Genetic Change: ______________________

____

____

Reading Check: What can cause gene flow? ______________________

____

SEXUAL REPRODUCTION AND EVOLUTION

Key Idea: Sexual reproduction creates the possibility that ______________ or ______________ can influence the gene pool of a population.

Additional notes about Sexual Reproduction and Evolution: ______________

____

____

POPULATION SIZE AND EVOLUTION

Key Idea: Allele frequencies are more likely to remain stable in

______________ populations than in ______________ populations.

Additional notes about Population Size and Evolution: ______________

____

____

Reading Check: What is the genetic effect of inbreeding? ______________

____

NATURAL SELECTION AND EVOLUTION

Key Idea: ______________________ acts only to change the relative frequency of alleles that exist in a population.

The word **deviate** means ______________________

____

Additional notes about How Selection Acts: ______________

____

____

Additional notes about Genetic Results of Selection: ______________

____

____

Reading Check: How is "fitness" measured in evolutionary terms? ______

____

Additional notes about Why Selection is Limited: ______________

____

____

Reading Check: How can unfavorable alleles persist? ______________

____

List possible exceptions to the statement "Natural selection removes unsuccessful phenotypes from a population." ______________________________

____

____

PATTERNS OF NATURAL SELECTION

Key Idea: Three major patterns are possible in the way that natural selection affects the distribution of polygenic characters over time: ______________________________

______________________________.

Additional notes about Directional Selection: ______________________________

____

____

Additional notes about Stabilizing Selection: ______________________________

____

____

Additional notes about Disruptive Selection: ______________________________

____

____

Reading Check: Which form of selection increases the range of variation in a distribution? ______________________________

____

Name ______________________ Class ______________ Date ______________

Note-taking Workbook

Population Genetics and Speciation

Section: Speciation

DEFINING SPECIES

Key Idea: Today, scientists may use more than one definition for ________.

The definition used depends on ________________ and ______________ ________________ being studied.

Additional notes about Defining Species: ________________________

____

____

Reading Check: Why is a species hard to define? ______________

____

FORMING NEW SPECIES

Key Idea: Speciation has occurred when the net effects of evolutionary forces result in a population that has ____________________________________

________________.

Reproductive isolation is ________________________________

____

A **subspecies** is ____

____

Additional notes about Reproductive Isolation: ________________

____

____

Additional notes about Mechanisms of Isolation: ______________

____

____

Reading Check: Is hybridization always successful? ____________

____

EXTINCTION: THE END OF SPECIES

Key Idea: The species that exist at any time are the next result of both

____________ and ____________.

Additional notes about Extinction: The End of Species: ____________

____

____

Reading Check: When do we know that extinction has happened?

____

____

Name ______________________ Class ______________ Date ______________

Note-taking Workbook

Classification

Section: The Importance of Classification

THE NEED FOR SYSTEMS

Key Idea: Biologists use ________________________________ to organize their knowledge of organisms. These ________________ attempt to provide consistent ways to name and categorize organisms.

Taxonomy is ____

____

Additional notes about The Need for Systems: ______________

____

____

Reading Check: What is the problem with common names of species? ___

____

____

SCIENTIFIC NOMENCLATURE

Key Idea: All scientific names for species are made up of two ___________ or ________________-like terms.

Genus is ____

____

Binomial Nomenclature is a ________________________________

____

Additional notes about Early Scientific Names: _____________

____

____

____

Additional notes about Naming Rules ______________________

____

____

____

Reading Check: Why did Linnaeus devise a new naming system? ______

____

____

THE LINNAEAN SYSTEM

Key Idea: In the Linnaean system of classification, organisms are grouped at successive levels of a hierarchy based on similarities in their ______________ and ______________.

Levels of the Linnaean System

Domain

____

Phylum

____

____

Family

____

Species

Additional notes about Levels of the Modern Linnaean System: ________

____

____

Reading Check: How many kingdoms are in the Linnaean system?

____

____

Name ______________________ Class ______________ Date ______________

Note-taking Workbook

Classification

Section: Modern Systematics

TRADITIONAL SYSTEMATICS

Key Idea: Scientists traditionally have used similarities in

______________ and ______________ to group organisms.

However, this approach has proven ______________.

Additional notes about Traditional Systematics: ______________

____

____

Reading Check: What is systematics? ______________

____

PHYLOGENETICS

Key Idea: Grouping organisms by ______________ is often assumed to reflect phylogeny, but inferring phylogeny is complex in practice.

Phylogeny is the ______________________________

____

Additional notes about Phylogenetics: ______________

____

____

____

CLADISTICS

Key Idea: Cladistic analysis is used to select the most likely ______________ among a given set of organisms.

Cladistics is a ______________________________

____

The word **objective** means ______________________

Additional notes about Cladistics: ______________________

____

____

The word root *clad* means "shoot, branch or twit" and the word root *gram* means "to write or record." Use this information to analyze the meaning of the term *cladogram.* ______________________

____

Reading Check: What does a cladogram show? ______________________

____

INFERRINGING EVOLUTIONARY RELATEDNESS

Key Idea: Biologists compare many kinds of ______________ and apply ______________ carefully in order to infer phylogenies.

Morphology refers to ______________________

____

Molecular evidence includes ______________________

____

Reading Check: What is an example of morphological data? ______________

____

The principle of parsimony holds that ______________________

____

Additional notes about Inferring Evolutionary Relatedness: ______________

____

____

Reading Check: What kinds of molecular data inform cladistics? ________

____

Name ______________________ Class ______________ Date ______________

Note-taking Workbook

Classification

Section: Kingdoms and Domains

UPDATING CLASSIFICATION SYSTEMS

Key Idea: Biologists have added ________________ and ____________ to classification systems as they have learned more.

Additional notes about Updating Classification Systems: ____________

____

____

Reading Check: What were the original Linnaean kingdoms? ________

____

THE THREE-DOMAIN SYSTEM

Key Idea: Today, most biologists tentatively recognize ________________ domains and ________________ kingdoms.

Bacteria are ____

____

Archaea are ____

____

A **eukaryote** is an ____

____

Major characteristics used to define kingdoms include:

cell type

____

____

body type

____

____

Additional notes about Major Characteristics: ____________

____

____

Additional notes about Domain Bacteria: ____________

____

____

Additional notes about Domain Archaea: ____________

____

____

The major groups of eukaryotes include:

Plantae

____

____

Protista

Additional notes about Domain Eukarya: ____________

____

____

Reading Check: Which kingdoms are prokaryotic? ____________

____

Reading Check: Which kingdoms are heterotrophic? ____________

____

Name ______________________ Class ______________ Date ______________

Note-taking Workbook

History of Life on Earth

Section: How Did Life Begin?

THE BASIC CHEMICALS OF LIFE

Key Idea: The ______________ experiment showed that, under certain conditions, ______________ compounds could form from ______________ molecules.

Additional notes about The Miller-Urey Experiment: ______________

____

____

Reading Check: What compounds were formed in the Miller-Urey experiment? ______________________________

____

LIFE'S BUILDING BLOCKS

Key Idea: Among the hypotheses that address the origin of life, one states that early ______________________ formed close to ______________ cents. Organic molecules may have also arrived on early Earth in ______________.

The word **impact** means ______________________________

____

Additional notes about Hydrothermal Vents: ______________

____

____

Additional notes about Space: ______________________

____

____

THE FIRST CELLS

Key Idea: Many scientists think that the formation of ______________

may have been the first step toward cellular organization.

A **microsphere** is a ______________________________

____

A **ribozyme** is a ______________________________

____

Additional notes about Forming a Cell: ______________________________

____

____

Additional notes about Origin of Heredity: ______________________________

____

____

Reading Check: Explain how RNA could have existed before DNA. ______

____

____

Name ______________________ Class ______________ Date ____________

Note-taking Workbook

History of Life on Earth

Section: The Age of Earth

THE FOSSIL RECORD

Key Idea: Both the geographical distribution of organisms and when they lived on Earth can be inferred from ______________________________, which chronicles the diversity of life on Earth.

A **fossil record** is ______________________________

____

Additional notes about How Fossils Form: ______________________________

____

____

ANALYZING FOSSIL EVIDENCE

Key Idea: In order to analyze fossil evidence, paleontologists use both ______________ and ______________ dating methods to date fossils.

Relative dating is a ______________________________

____

Radiometric dating is a ______________________________

____

Half-life is ______________________________

____

Additional notes about Types of Fossils: ______________________________

____

____

Additional notes about Relative Age: ______________________________

____

____

Additional notes about Absolute Age: ________________________________

____

____

Reading Check: What is the law of superposition? ______________

____

DESCRIBING GEOLOGIC TIME

Key Idea: The geologic time scale is based on evidence in the ______________ ______________ and has been shaped by mass ______________.

The **geologic time scale** is ________________________________

____

Mass extinction is ________________________________

____

Earth's history is divided into three eras:

Paleozoic Era ________________________________

Mesozoic Era ________________________________

Cenozoic Era ________________________________

Additional notes about Divisions of Geologic Time: ______________

____

____

Additional notes about Mass Extinction: ______________________

____

____

Reading Check: What evidence shows that mass extinctions occur?

____

____

Name ______________________ Class ____________ Date ____________

Note-taking Workbook

History of Life on Earth

Section: Evolution of Life

PRECAMBRIAN TIME

Key Idea: Single-celled ________________ and later, ________________, evolved and flourished in ________________ time. The evolution of ________________ set the stage for the evolution of modern organisms. The accumulation of ________________ allowed organisms to live on land.

Cyanobacteria are ______________________________

____

Endosymbiosis is a ______________________________

____

The word **accumulate** means ______________________________

Additional notes about Prokaryotic Life: ______________________________

____

____

Additional notes about Formation of Oxygen: ______________________________

____

____

Additional notes about Eukaryotic Life: ______________________________

____

Observations that support the theory of endosymbiosis include:

Size and Structure: ______________________________

Genetic Material: ______________________________

Ribosomes: ______________________________

Reproduction: ______________________________

Additional notes about Origin of Energy-Producing Organelles: ______

____

____

Additional notes about Multicellularity: ______________

____

____

Additional notes about Dominant Life: ______________

____

____

Additional notes about Mass Extinctions: ______________

____

____

Reading Check: Why is the evolution of colonial organisms an important step in evolution? ______________

____

PALEOZOIC ERA

Key Idea: During the Paleozoic Era, marine ______________ diversified, and marine ______________ evolved. The first ______________ evolved. Some ______________, and then some ______________, left the oceans to colonize land.

Additional notes about Dominant Life: ______________

____

____

Additional notes about Mass Extinctions: ______________

____

____

MESOZOIC AND CENOZOIC ERAS

Key Idea: ________________, ________________, and ____________

were the dominant animals during the ________________ Era, and

__________________________________ dominated the

________________ Era.

Additional notes about Dominant Life: ________________________

____

____

Additional notes about Mass Extinction: ______________________

____

____

Name ______________________ Class ______________ Date ____________

Note-taking Workbook

Bacteria and Viruses

Section: Bacteria

WHAT ARE PROKARYOTES?

Key Idea: Prokaryotes are divided into two major groups: the domain ______________ and the domain ______________.

Additional notes about Archaea: ________________________________

____

____

Additional notes about Bacteria: ________________________________

____

____

BACTERIAL STRUCTURE

Key Idea: ______________________ bacteria have a thick layer of ______________ and no outer membrane. ______________ ______________ bacteria have a thin layer of ______________ and have an outer membrane.

A **plasmid** is a ____

____

Peptidoglycan is a ____

____

Gram-positive is a ____

____

Gram-negative is a ____

____

Additional notes about Gram-Positive Bacteria: ______________________

____

____

Additional notes about Gram-Negative Bacteria: ______________________

____

____

Reading Check: Is *E. coli* a Gram-positive or Gram-negative bacterium?

____

____

OBTAINING ENERGY AND NUTRIENTS

Key Idea: Grouping prokaryotes based on their energy source separates them into ______________, ______________, and ______________.

Additional notes about Photoautotrophs: ______________________

____

____

Additional notes about Chemoautotrophs: ______________________

____

____

Additional notes about Heterotrophs: ______________________

____

REPRODUCTION AND ADAPTATION

Key Idea: Prokaryotes reproduce by binary fission; ______________

______________________________, ______________,

and______________; and survive harsh conditions by forming

______________ .

Conjugation is ____

____

Transformation is ____

____

Transduction is ____

____

Endospore is ____

____

Additional notes about Binary Fission: ____________________________

____

____

Additional notes about Genetic Recombination: ____________________

____

____

Additional notes about Endospore Formation: ______________________

____

____

Name ______________________ Class ______________ Date ______________

Note-taking Workbook

Bacteria and Viruses

Section: Viruses

IS A VIRUS ALIVE?

Key Idea: Viruses are ________________ living because they ____________ key characteristics of living organisms.

Additional notes about Is a Virus Alive?: ______________________

____

____

VIRAL STRUCTURE

Key Idea: All viruses have ____________________ and a __________.

A **capsid** is a ____

____

An **envelope** is a ____

____

A **bacteriophage** is a ____

____

Additional notes about Nucleic Acids: ______________________

____

____

Additional notes about Capsid: ______________________

____

____

Additional notes about Envelope: ______________________

____

____

Additional notes about Tail Fibers: ______________

____

____

Reading Check: How does reproduction differ between DNA and RNA viruses? ______________

____

REPRODUCTION

Key Idea: Viruses can reproduce by a ______________ life cycle and a ______________ life cycle.

Lytic is a ______________

____

Lysogenic is a ______________

____

Additional notes about Lytic Cycle: ______________

____

Additional notes about Lysogenic Cycle: ______________

____

____

VIROIDS AND PRIONS

Key Idea: Viroids and prions are molecules that are able to ______________ and ______________.

Additional notes about Viroids: ______________

____

Additional notes about Prions: ______________

____

Name ______________________ Class ______________ Date ______________

Note-taking Workbook

Bacteria and Viruses

Section: Bacteria, Viruses, and Humans

ROLES OF BACTERIA AND VIRUSES

Key Idea: Bacteria play important roles in the ______________ and in ______________. Both bacteria and viruses are important in research.

Additional notes about Bacteria and the Environment: ______________

____

____

Additional notes about Bacteria and Industry: ______________

____

____

Additional notes about Bacteria, Viruses, and Research: ______________

____

____

KOCH'S POSTULATES AND DISEASE TRANSMISSION

Key Idea: The four main steps in Koch's postulates are ______________

____

____.

Koch's postulates is a ______________________________

____

A **pathogen** is an ______________________________

____

Additional notes about Koch's Postulates and Disease Transmission:

____

____

____

Reading Check: What are five ways diseases can be transmitted? ______

____

BACTERIAL DISEASES

Key Idea: Bacteria can cause disease by producing ______________ and by ______________________________.

A **toxin** is a ______________________________

____

Additional notes about Bacterial Diseases: ______________

____

____

Use the two-column table below. In the "Effect" column, list all of the diseases discussed in this section. In the "Cause" column, list the pathogen that causes the disease, and note whether the pathogen is a bacterium or a virus.

Cause	Effect

ANTIBIOTIC RESISTANCE

Key Idea: Antibiotic resistance spreads when sensitive populations of ______________ are killed by ______________. As a result, resistant bacteria ______________.

An **antibiotic** is a ____

____

Resistance is ____

____

The word **effective** means ______________________________

Additional notes about Development of Resistance: ____________

____

Additional notes about Consequences of Resistance: ____________

____

____

VIRAL DISEASES

Key Idea: Because viruses enter ______________ to reproduce, it is difficult to develop a drug that kills the virus without harming the living host.

Additional notes about Viral Diseases: ____________________

____

____

Reading Check: What factors cause the symptoms of viral disease?

____

____

EMERGING DISEASES

Key Idea: Emerging diseases are infectious diseases that are ____________ ______________________________, that have ______________ ______________________________, or that have ______________ when a disease that was once considered under control begins to spread.

Additional notes about Emerging Diseases: ______________

____

____

Name ______________________ Class ______________ Date ____________

Note-taking Workbook

Protists

Section: Characteristics of Protists

WHAT ARE PROTISTS?

Key Idea: Protists are ________________ organisms that cannot be classified as ______________, ________________ or animals.

Additional notes about What Are Protists?: ________________________

____

____

Reading Check: what important characteristics arose among protests during their evolution? ______________________________

____

REPPRODUCTION

Key Idea: Protists can reproduce asexually by ________________, ________________, and ________________. Protists can also reproduce sexually by ________________ of ________________.

A **gamete** is a ______________________________________

____

Zygote is ______________________________________

____

Zygospore is ______________________________________

____

The **alterations of generations** is ______________________________

____

Binary Fission: ______________________________________

Budding: ______________________________________

Fragmentation: ______________________________________

Additional notes about Asexual Reproduction: ______________

____

____

Additional notes about Sexual Reproduction: ______________

____

____

Reading Check: How does alternation of generations differ from sexual reproduction in unicellular protists? ______________

____

The Greek word root *phyte* means "plant." Using this information, propose your own definitions for *sporophyte* and *gametophyte*. ______________

____

CLASSIFYING PROTISTS

Key Idea: The classification of organisms currently grouped in ______________ ______________ is likely to ______________ as scientists learn more about ______________ and ______________

____.

Additional notes about Classifying Protists: ______________

____

____

Name ______________________ Class ______________ Date ____________

Note-taking Workbook

Protists

Section: Groups of Protists

GROUPING PROTISTS

Key Idea: Grouping protists by the way they ______________________

______________________ helps us understand their ecological roles.

Additional notes about Grouping Protists: ______________________

____

____

Reading Check: What method can be used to group protists? __________

____

ANIMAL-LIKE PROTISTS

Key Idea: Animal-like protists ______________ other organisms to

______________________________.

A **pseudopodium** is a ______________________________

____

The word **variety** means ______________________________

____

Additional notes about Amoeboid Protists: ______________________

____

____

Additional notes about Ciliates: ______________________________

____

____

Additional notes about Flagellates: ______________________________

____

____

Additional notes about Sporozoans: ______________________

____

____

Reading Check: Which group of protists is all parasitic? ______________

____

PLANTLIKE PROTISTS

Key Idea: Plantlike protists obtain energy through ______________.

Additional notes about Diatoms: ______________________

____

____

Additional notes about Euglenoids: ______________________

____

____

Additional notes about Dinoflagellates: ______________________

____

____

Reading Check: In which group of protists do the individuals get smaller every time they reproduce asexually? ______________

____

Write a general statement that describes plantlike protists. ______________

____

Find two exceptions to this general statement. ______________

____

____

____

Additional notes about Red Algae: ______________________

____

____

Additional notes about Brown Algae: ______________________

____

____

Additional notes about Green Algae: ______________________

____

____

FUNGUSLIKE PROTISTS

Key Idea: ______________ protists absorb ______________ from their environment and reproduce by ______________.

A **plasmodium** is ______________________

____

Additional notes about Slime Molds: ______________________

____

____

Additional notes about Water Molds and Downy Mildews: ______________

____

____

Name ______________________ Class ______________ Date ______________

Note-taking Workbook

Protists

Section: Protists and Humans

PROTISTS AND DISEASE

Key Idea: Protists ________________ a number of human diseases, including giardiasis, amebiasis, toxoplasmosis, trichomoniasis, cryptosporidiosis, Chagas disease, and malaria.

The word **rarely** means ______________________________

____

Giardiasis

Cause: ______________________________

Symptoms: ______________________________

Amebic Dysentery

Cause: ______________________________

Symptoms: ______________________________

Toxoplasmosis

Cause: ______________________________

Symptoms: ______________________________

Trichomoniasis

Cause: ______________________________

Symptoms: ______________________________

Cryptosporidiosis

Cause: ______________________________

Symptoms: ______________________________

Chagas Disease

Cause: ______________________________

Symptoms: ______________________________

Malaria

Cause: ______________________________

Symptoms: ______________________________

Additional notes about Protists and Disease: ______________________________

____

PROTISTS AND THE ENVIRONMENT

Key Idea: Protists produce ______________, take up ______________, are important producers in ______________ food webs, can produce ______________________, serve as ______________ ______________ and have ______________ relationships with many animals and plants.

An **algal bloom** is a ______________________________

____

Additional notes about Protists and the Environment: ______________

____

____

____

Reading Check: What are three ways in which protists affect ocean ecosystems? ______________________________

____

PROTISTS AND INDUSTRY

Key Idea: Protists are important in many ______________, in industrial and consumer ______________, and in ______________.

Additional notes about Protists and Industry: ______________________________

____

____

Name ______________________ Class ______________ Date ______________

Note-taking Workbook

Fungi

Section: Characteristics of Fungi

WHAT ARE FUNGI?

Key Idea: Fungi have ______________ bodies, their cell walls are made of ______________, and they absorb ______________.

A **chitin** is ______________________________

Additional notes about What Are Fungi?: ______________________________

STRUCTURE AND FUNCTION

Key Idea: A typical fungal body is made of ______________ that allow the fungus to have a ______________ surface area and to absorb nutrients efficiently.

A **hypha** is ______________________________

The **mycelium** is the ______________________________

A **rhizoid** is a ______________________________

A **saprobe** is an ______________________________

Additional notes about Body Structure: ______________________________

Additional notes about Obtaining Nutrients: ______________________

____

____

REPRODUCTION

Key Idea: In sexual reproduction, ______________ are produced by ______________. In asexual production, ______________ are produced by ______________.

Additional notes about Sexual Reproduction: ______________________

____

____

Additional notes about Asexual Reproduction: ______________________

____

____

Additional notes about Yeast and Mold: ______________________

____

____

Reading Check: What is the difference between spores produced sexually and spores produced asexually in fungi? ______________________

____

Write two sentences that compare and two sentences that contrast sexually and asexually produced by spores. ______________________

____

____

Name ______________________ Class ______________ Date ______________

Note-taking Workbook

Fungi

Section: Groups of Fungi

CHYTRID FUNGI

Key Idea: The ________________ are a group of ________________ fungi that provide clues about ______________________________.

Additional notes about Chytrid Fungi: ______________________________

____

____

Reading Check: Which characteristics do chytrids share with protists, and which do they share with other fungi? ______________________

____

ZYGOTE FUNGI

Key Idea: Zygote fungi are named for ______________________________ that produce ________________ inside a tough capsule.

A **zygosporangium** is a ______________________________

____

The word **identical** means ______________________________

____

Additional notes about Zygote Fungi: ______________________________

____

____

Reading Check: Where does meiosis take place in zygote fungi?

____

____

SAC FUNGI

Key Idea: ________________ are characterized by an ascus, a ____________ ____________________________ that produces spores.

An **ascus** is ____

Additional notes about Sac Fungi: ______________________________

Reading Check: In sac fungi, which structure is dikaryotic? ________

Fill in the process chart below with the steps of the life cycle of sac fungi. Label the loops for sexual and asexual reproduction.

1: Hyphae from two mating types fuse.

2:

3: Inside the ascocarp asci form Nuclei within the asci fuse.

4:

5:

6: Spores germinate to form new fungal hyphae.

7:

CLUB FUNGI

Key Idea: Club fungi are characterized by a ________________, a clublike sexual reproductive structure that produces spores.

A **basidium** is ____

____

Additional notes about Club Fungi: ______________________

____

____

Reading Check: Which part of a club fungus is dikaryotic? __________

____

FUNGAL PARTNERSHIPS

Key Idea: Fungi from mutualistic symbiotic associations to form __________ and ________________.

A **lichen** is a ____

____

A **mycorrhiza** is ______________________________________

____

Additional notes about Fungal Partnerships: ______________

____

____

Name ______________________ Class ______________ Date ______________

Note-taking Workbook

Fungi

Section: Fungi and Humans

FUNGI AND INDUSTRY

Key Idea: Fungi are used for ______________, ______________, ______________, ______________, and ________.

Additional notes about Fungi and Industry: ______________

____

____

FUNGI AND THE ECOSYSTEM

Key Idea: Fungi play important ecological roles by ______________ ____.

Additional notes about Fungi and the Ecosystem: ______________

____

____

Reading Check: What is the primary role of fungi in ecosystems? ________

____

____

FUNGI AND DISEASE

Key Idea: Fungi cause disease by ______________ nutrients from host ______________ and by ______________________________.

A **dermatophyte** is ____

____

Additional notes about Fungal Infections: ______________

____

____

Additional notes about Fungal Toxins: ______________________

____

____

The prefix *histo-* means "a web." Explain why this prefix might be art of an appropriate name for a fungal disease. ______________________

____

____

Name ______________________ Class ______________ Date ________

Note-taking Workbook

Plant Diversity and Life Cycles

Section: Introduction to Plants

WHAT IS A PLANT?

Key Idea: Plants are ______________________ whose cells have cell walls. Most plants are ______________ - they produce their own food through ______________.

Additional notes about What is a Plant?: ______________________

____

____

Reading Check: What do plants need for photosynthesis? ____________

____

ESTABLISHMENT OF PLANTS ON LAND

Key Idea: In order to thrive on land, plants had to be able to absorb ______________________, to ______________________, and to ______________________.

A **cuticle** is a ______________________________________

____

A **spore** is a ______________________________________

____

The word **transport** means ______________________________

Additional notes about Absorbing Nutrients: ____________________

____

____

Additional notes about Preventing Water Loss: __________________

____

____

Additional notes about Dispersal on Land: ________________________

____

____

Reading Check: What is the waxy layer on the aboveground parts of most plants that helps prevent water loss called? ________________________

____

PLANT LIFE CYCLES

Key Idea: Plants have life cycles in which ________________ alternate with ________________________. A life cycle in which a ________________________ alternates with a ________________ is called ________________________.

A **sporophyte** is ________________________________

____

A **gametophyte** is ________________________________

____

Basic Life Cycle of a Plant

Step 1: ________________________________

Step 2: ________________________________

Step 3: ________________________________

Additional notes about Plant Life Cycles: ________________________

____

____

Reading Check: Is a plant sporophyte diploid or haploid? ____________

____

Name ______________________ Class ______________ Date ______________

Note-taking Workbook

Plant Diversity and Life Cycles

Section: Seedless Plants

NONVASCULAR PLANTS

Key Idea: Nonvascular plants are small plants that reproduce by means of ________________. They lack true ________________, ________________, and ________________, which are complex structures that contain vascular, or conducting, tissues.

The word **consist** means ____

____

Additional notes about Mosses: ____

____

Additional notes about Liverworts: ____

____

Additional notes about Hornworts: ____

____

REPRODUCTION IN NONVASCULAR PLANTS

Key Idea: In the life cycle of nonvascular plants, the ________________ is the dominant generation. ________________ must be covered by a film of ________________ in order for ________________ to occur.

An **archegonium** is a ____

____

An **antheridium** is a ____

____

A **sporangium** is a ____

____

Additional notes about Life Cycle of a Moss: ____________________

____

____

Reading Check: Which structure produces male sex cells in nonvascular plants? ____________________

____

SEEDLESS VASCULAR PLANTS

Key Idea: ______________ of seedless vascular plants have vascular tissue, but ______________ lack vascular tissue. Because of their vascular system, vascular plants grow much ______________ than nonvascular plants and also develop ______________.

A **rhizome** is a ______________________________

____

A **frond** is the ______________________________

____

Additional notes about Club Mosses: ____________________

____

____

Additional notes about Ferns and Fern Allies: ____________________

____

____

Reading Check: Where do sporangia form on ferns? ____________________

____

What are two types of seedless vascular plants? ____________________

____

REPRODUCTION IN SEEDLESS VASCULAR PLANTS

Key Idea: Like nonvascular plants, seedless vascular plants can reproduce

________________ when a film of water covers the ________________.

Unlike nonvascular plants, seedless vascular plants have ________________

that are much larger than their ________________.

A **sorus** is a ____

____

Additional notes about Reproduction in Seedless Vascular Plants:

____

____

____

Reading Check: How large is a fern gametophyte? ________________

____

Additional notes about Spores: ________________________________

____

____

Reading Check: What is a cluster of sporangia on a fern frond called?

____

____

Name ______________________ Class ______________ Date ______________

Note-taking Workbook

Plant Diversity and Life Cycles

Section: Seed Plants

KINDS OF SEED PLANTS

Key Idea: Seed plants are traditionally classified into two groups - ________________ and ________________.

A **gymnosperm** is ____
____

A **angiosperm** is ____
____

Additional notes about Kinds of Seed Plants: ______________________
____
____

Reading Check: What is the difference between gymnosperms and angiosperms in terms of seed production? ______________________
____

REPRODUCTION IN SEED PLANTS

Key Idea: Unlike seedless plants, seed plants do not require ____________ to reproduce sexually. Reproduction in seed plants is also characterized by a greatly reduced ________________ and a dominant ________________.

A **ovule** is a ____
____

A **seed** is a ____
____

The **pollen grain** is the ______________________________________
____

Pollination is the ____

____

Additional notes about Pollination and Fertilization: ______________

____

____

Additional notes about Seed Formation: ______________________

____

____

Additional notes about Seed Dispersal: ______________________

____

____

Reading Check: Where do gametophytes develop in seed plants?

____

Fill in this concept map using what you have learned about reproduction in seed plants.

Plant reproduction

can be

sexual

which involves

[ ] in the [ ] during pollination

[ ] in pollen transferred to [ ]

GYMNOSPERMS

Key Idea: There are four major groups of gymnosperms - ______________

____.

Additional notes about Gymnosperms: ______________

____

____

____

Reading Check: Which gymnosperm has seeds that do not develop within a cone? ______________________________

LIFE CYCLE OF A CONIFER

Key Idea: Reproduction in conifers is characterized by a dominant ______________, ______________________, and the development of ______________________.

The word **cycle** means ______________________

Additional notes about Life Cycle of a Conifer: ______________

____

____

Reading Check: What characteristics of pine seeds aids in dispersal of the seeds? ______________________________

Additional notes about Cones: ______________________

____

____

Reading Check: Are male and female cones always produced on separate plants? ______________________________

Note-taking Workbook

Plant Diversity and Life Cycles

Section: Flowering Plants

KINDS OF ANGIOSPERMS

Key Idea: Botanists traditionally divide the angiosperms into two subgroups –
____.

A **monocot** is an ____
____

A **cotyledon** is ____
____

A **dicot** is an ____
____

Additional notes about Kinds of Angiosperms: ______________________
____
____

Reading Check: What are three characteristics of monocots?
____
____

REPRODUCTION IN ANGIOSPERMS

Key Idea: A ________________ is a specialized reproductive structure of angiosperms. The male and female ________________ of angiosperms develop within ________________, which promote ________________ and ________________ more efficiently than do cones.

A **stamen** is ____
____

An **anther** is ____
____

A **pistil** is ____

____

Additional notes about Structure of Flowers: ______________

____

____

Additional notes about Kinds of Flowers: ______________

____

Reading Check: What is the function of a stamen? ______________

____

Additional notes about Life Cycle of an Angiosperm: ______________

____

____

Reading Check: What is the function of a pollen tube? ______________

____

POLLINATION

Key Idea: The flowers of many ______________ are adapted for pollination by ______________ or by ______________.

Additional notes about Life Cycle of an Angiosperm: ______________

____

____

Reading Check: Name three characteristics of flowers that might attract pollinators. ____

____

List two characteristics of insect-pollinated flowers, then list two characteristics of wind-pollinated flowers. ____

____

FRUITS

Key Idea: Although fruits provide some protection for developing seeds, they primarily function in ____________________.

Fruit is ____

____

Additional notes about Fruits: ____________________

____

____

Reading Check: From which part of a flower does a fruit develop?

____

VEGATATIVE REPRODUCTION

Key Idea: Plants reproduce ______________ in a variety ways that involve ______________ parts, such as ______________, ______________, and ______________. The reproduction of plants from these parts is called ____________________.

Additional notes about Vegetative Reproduction: ______________

____

____

Reading Check: What are three types of modified stems by which plants can reproduce vegetatively? ____________________

____

Name ______________________ Class ______________ Date ______________

Note-taking Workbook

Seed Plant Structure and Growth

Section: Plant Tissue Systems

PLANT TISSUES

Key Idea: Vascular plants have three tissue systems – ______________________

____.

Dermal tissue is ____

____

Vascular tissue is ____

____

Ground tissue is ____

____

Additional notes about Plant Tissues: __________________________

____

____

Reading Check: Where on a plant is dermal tissue found? __________

____

DERMAL TISSUE SYSTEM

Key Idea: Dermal tissue covers the ______________ of a plant's body. In the ______________ parts of a plant, dermal tissue forms a "skin" called the ______________.

A **stoma** is the ____

____

A **guard cell** is ____

____

The word **function** means ______________________________________

____

Additional notes about Dermal Tissue System: ______________________

____

____

Reading Check: What is the function of root hairs? ______________

____

VASCULAR TISSUE SYSTEM

Key Idea: Vascular plants have two kinds of vascular tissue, called

______________ and ______________, that transport ______________,

______________ and ______________ throughout the plant.

Xylem is the ______________________________________

____

Phloem is ______________________________________

____

Additional notes about Vascular Tissue System: ______________________

____

____

Reading Check: What are the conducting cells in phloem called?

____

GROUND TISSUE SYSTEM

Key Idea: Ground tissue makes up much of the ______________________,

where it ______________ and ______________ vascular tissue.

Additional notes about Ground Tissue System: ______________________

____

____

Reading Check: What is the primary function of ground tissue in roots and

stems? ____

Name ______________________ Class ______________ Date ____________

Note-taking Workbook

Seed Plant Structure and Growth

Section: Roots, Stems, and Leaves

ROOTS

Key Idea: Most plants are anchored to the spot where they grow by roots, which absorb ______________ and ______________. In many plants, roots also function in the storage of organic ______________, such as ______________ and ______________.

Additional notes about Roots: ______________

____

____

Reading Check: What are three functions of roots? ______________

____

STEMS

Key Idea: Stems support the leaves and house the ______________ tissue, which transports substances between the roots and the ______________.

A **vascular bundle** is a ______________

____

A **pith** is the ______________

____

Heartwood is the ______________

____

Sapwood is the ______________

____

Additional notes about Nonwoody Stems: ______________

____

____

Additional notes about Woody Stems: ______________________

____

____

Reading Check: What is the name of the point where a leaf attaches to a stem?

____

____

LEAVES

Key Idea: Leaves are the primary ____________ organs of plants.

A **blade** is the ________________________

____

A **petiole** is the ________________________

____

Mesophyll is ________________________

____

The word **source** means ________________________

____

Additional notes about Specialized Leaves: ______________

____

____

Reading Check: What are two types of specialized leaves, and what are their functions? ________________________

____

Seed Plant Structure and Growth

Section: Plant Growth and Development

THE PLANT EMBRYO

Key Idea: The plant embryo possesses an embryonic ______________________ and an embryonic ______________________. Leaflike structures called ______________________, or seed leaves, are attached to the embryonic ______________________.

Germination is ____

____

Additional notes about The Plant Embryo: ________________________________

____

____

Reading Check: How many cotyledons does a bean seed have?

____

____

Additional notes about Germination: ________________________________

____

Additional notes about Breaking Dormancy: ________________________________

____

Reading Check: What are two ways in which the seed coat can be damaged so that the seed will be able to sprout? ________________________________

____

Use cause and effect language to describe what happens when water penetrates a seed coat. ____

____

MERISTEMS

Key Idea: Plants grow by producing new ________________ in regions of active cell division called ________________.

A **meristem** is a ____
____

Primary growth is ____
____

Secondary growth is ____
____

Additional notes about Meristems: ____________________________
____
____

Reading Check: How does primary growth differ from secondary growth?
____
____

PRIMARY GROWTH

Key Idea: Primary growth makes a plant's stems and roots get ______________ without becoming ________________.

Apical meristem is ____
____

Additional notes about Primary Growth: ______________________
____
____

Reading Check: How many apical meristems does a plant embryo have?
____
____

SECONDARY GROWTH

Key Idea: Lateral meristems are responsible for increases in the

________________ of stems and roots. This increase is called

________________________________.

Lateral meristem is ______________________________________

____

Additional notes about Secondary Growth: ____________________

____

____

Reading Check: What are the names of the two lateral meristems that are

responsible for secondary growth? ____________________________

____

Name ____________________ Class ____________ Date ____________

Note-taking Workbook

Plant Processes

Section: Nutrients and Transport

NUTRIENTS

Key Idea: ________________, ________________, and ________________ do not satisfy all of a plant's need for raw materials.

Plants also require small amounts of ____________________, which are elements absorbed mainly as ____________________.

Additional notes about Nutrients: ____________________

____

____

Reading Check: What two raw materials do plants need to make carbohydrates? ____________________

____

TRANSPORT OF WATER

Key Idea: Water and mineral nutrients more up from a plant's roots to its leave through ________________.

Transpiration is ____________________

____

Additional notes about Transport of Water: ____________________

____

Reading Check: What causes the upward pull on water molecules in xylem? ____________________

____

Use spatial language to describe the transport of water through a plant. ______

____

____

Additional notes about Guard Cells and Transpiration: ______________

____

____

Reading Check: What happens to the guard cells and stoma when the guard cells take in water? ______________________________

____

TRANSPORT OF ORGANIC COMPOUNDS

Key Idea: Organic compounds move through a plant within the

______________ from a ______________ to a ______________.

Steps in the Pressure-Flow Model

Step 1: ____

Step 2: ____

Step 3: ____

Step 4: ____

Additional notes about Pressure-Flow Model: ______________

____

____

Reading Check: What is an example of a sink? ______________

____

Name ______________________ Class ______________ Date ______________

Note-taking Workbook

Plant Processes

Section: Plant Responses

PLANT HORMONES

Key Idea: Plant hormones are produced in small amounts buy may have large effects on the growth and development of plants. Hormones may ______________ or ______________ growth in a plant.

Reading Check: What are two ways in which hormones can affect the growth and development of a plant? ______________________________

____

Steps in Went's experiment

Step 1: ______________________________

Step 2: ______________________________

Step 3: ______________________________

Step 4: ______________________________

Additional notes about Auxins: ______________________________

____

Reading Check: What happened to the oat shoot when the agar block with auxin was applied to it? ______________________________

Additional notes about Gibberellins: ______________________________

____

Additional notes about Cytokinins: ______________________________

____

Additional notes about Ethylene: ______________________________

____

Additional notes about Abscisic Acid: ______________________________

____

Reading Check: What is the one effect of each of the plant hormones discussed in this section? ______________________

____

TROPISMS

Key Idea: A tropism is a response in which a plant grows toward or away from a ______________. Plant ______________ are responsible for producing tropisms.

A **tropism** is ______________________

____

A **phototropism** is ______________________

____

A **thigmotropism** is ______________________

____

A **gravitropism** is ______________________

____

Additional notes about Phototropism: ______________________

____

Additional notes about Thigmotropism: ______________________

____

Additional notes about Gravitropism: ______________________

____

Reading Check: What is a negative tropism? ______________________

____

Use what you know about word parts to write a definition of the word *heliotropism.* ______________________

____

SEASONAL RESPONSES

Key Idea: The principal way in which plants time seasonal responses is by sensing changes in _______________ length.

Photoperiodism is ______________________________

____

Dormancy is ______________________________

____

The word **categorize** means ______________________________

____

Additional notes about Photoperiodism: ______________________________

____

Additional notes about Responses to Temperature: ______________________________

____

Reading Check: What are plants whose growth is not affected by day length called? ______________________________

____

PLANT MOVEMENTS

Key Idea: Some plant movements respond to an _______________ stimulus but are not influenced by the _______________ of the stimulus.

Nastic movement is ______________________________

____

Additional notes about Plant Movements: ______________________________

____

Reading Check: How are nastic movements regulated at the cellular level?

____

____

Name ______________________ Class ______________ Date ______________

Note-taking Workbook

Introduction to Animals

Section: Characteristics of Animals

GENERAL FEATURES OF ANIMALS

Key Idea: Animals are ______________, ______________ organisms with cells that lack cell ______________.

A **heterotroph** is ______________________________

____

The word **environment** means ______________________________

____

Additional notes about General Features of Animals: ______________

____

Reading Check: List three ways in which humans depend on animals.

____

____

Additional notes about Multicellularity: ______________

____

Additional notes about Heterotrophy: ______________

____

Additional notes about Movement: ______________

____

Reading Check: What are three advantages of being able to move around the environment? ______________

____

KINDS OF ANIMALS

Key Idea: Animals are often informally grouped as ________________ or ________________ although ________________ make up only a subgroup of one phylum - ________________. The vast majority of animals are ________________.

An **invertebrate** is ____
____

A **vertebrate** is ____
____

Additional notes about Invertebrates: ____
____
____

Reading Check: Explain why many invertebrates are small. ____________
____

Additional notes about Vertebrates: ____
____
____

Reading Check: What are two characteristics that all vertebrates share?
____
____

Name ______________________________ Class __________________ Date ________________

Note-taking Workbook

Introduction to Animals

Section: Animal Body Systems

SUPPORT

Key Idea: An animal's ____________________ provides a framework that supports the animal's body. The __________________ is also vital to an animal's movement.

A **hydrostatic skeleton** is ____

____

An **exoskeleton** is ____

____

An **endoskeleton** is ____

____

Additional notes about Support: ______________________________

____

____

Reading Check: What are three types of skeletons? _____________

____

DIGESTIVE AND EXCRETORY SYSTEMS

Key Idea: The digestive system is responsible for _________________ energy and nutrients from an animal's food, while the excretory system ____________________ waste products from the animal's body.

The **gastrovascular cavity** is __________________________________

____

Additional notes about Digestive System: ______________________

____

____

Additional notes about Excretory System: ______________________

____

____

Reading Check: How does a gastrovascular cavity of a hydra differ from a one-way digestive system? ______________________

____

NERVOUS SYSTEM

Key Idea: The nervous system carries ______________________ through the body and coordinates ______________ and ______________.

Additional notes about Nervous Systems: ______________________

____

____

____

Reading Check: What type of nervous system do jellyfish have?

____

Use spatial language to describe the nervous systems of a hydra, a flatworm, and a grashopper. ______________________

____

____

RESPIRATORY AND CIRCULATORY SYSTEMS

Key Idea: The respiratory system is responsible for ______________________ between the body and the environment. The circulatory system transports ______________________ within the body.

The word **transports** means ______________________

____

Additional notes about Respiratory System: ______________________

____

____

Additional notes about Circulatory System: ______________________

____

____

Reading Check: What is the function of the circulatory system?

____

____

REPRODUCTION

Key Idea: The two types of reproduction in animals are ________________

and ________________.

Additional notes about Asexual Reproduction: ______________________

____

____

Additional notes about Sexual Reproduction: ______________________

____

____

Reading Check: Name an animal that can reproduce asexually.

____

____

Name ______________________ Class ______________ Date ______________

Note-taking Workbook

Introduction to Animals

Section: Evolutionary Trends in Animals

TISSUES AND SYMMETRY

Key Idea: Through evolutionary time, animals have developed more

______________ body plans, including ______________________

and ______________________.

Cephalization is ______________________

____

Additional notes about Tissues: ______________________

____

____

Additional notes about Body Symmetry: ______________________

____

____

Reading Check: What are tissues, and what is an example of a type of

tissue? ______________________

____

EARLY EMBRYONIC DEVELOPMENT

Key Idea: The ______________ that is produced through the union of

______________ and ______________ undergoes cell division and

tissue development during ______________, ______________,

and ______________.

Cleavage is ______________________

____

The **blastula** is ______________________

____

Gastrulation is ____

____

A **protostome** is an ____

____

A **deuterostome** is ____

____

Additional notes about Differentiation: ______________________________

____

____

Reading Check: What are the three primary tissue layers? ______________

____

Additional notes about Patterns of Development: ______________

____

____

Reading Check: Where does the mouth form in protostomes?

____

Fill in the spider map below to help you organize the information you learned about early embryonic development in animals.

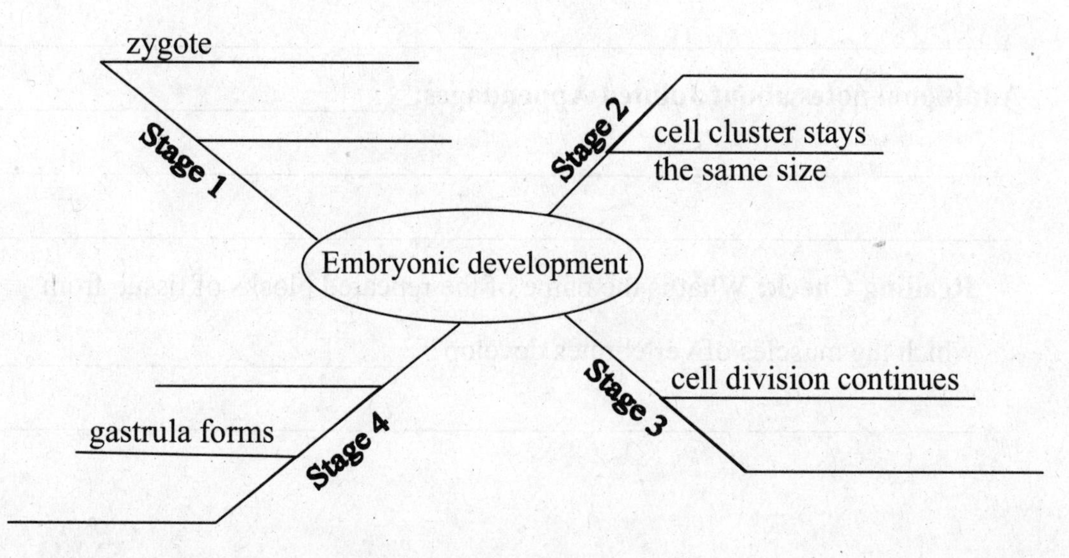

BODY CAVITIES

Key Idea: Animals with ________________ symmetry have one of three basic kinds of internal body plans. The body plan may include a body cavity, or ________________.

A **coelom** is a ____

____

Additional notes about A True Coelom: ________________

____

____

Reading Check: What is an example of a psedocoelomate? ________________

____

SEGMENTATION AND JOINTED APPENDAGES

Key Idea: Two major body characteristics evolved that gave animals a greater ability to move and to be flexible. These two evolutionary trends in animals are ________________ and ________________ ________________.

Additional notes about Body Segmentation: ________________

____

____

Additional notes about Jointed Appendages: ________________

____

____

Reading Check: What is the name of the repeated blocks of tissue from which the muscles of vertebrates develop? ________________

____

Name ______________________ Class ______________ Date ______________

Note-taking Workbook

Introduction to Animals

Section: Chordate Evolution

CHARACTERISTICS OF CHORDATES

Key Idea: At some point in their development, all chordates have a

______________ nerve cord, a ______________, ______________

______________ and a ______________.

A **notochord** is the ______________

____

Additional notes about Invertebrate Chordates: ______________

____

____

Additional notes about The First Vertebrates: ______________

____

____

Reading Check: What are two types of invertebrate chordates?

____

____

EVOLUTION OF FISHES

Key Idea: Two important structures that first evolved in fish allowed them to

become efficient underwater predators. ______________ and

______________ allowed fish to pursue and grasp prey.

Additional notes about Transitional Forms: ______________

____

____

____

EVOLUTION OF AMPHIBIANS

Key Idea: Three major characteristics helped amphibians succeed on land.

First, amphibians had ____
____.

Second, an amphibian's ____
____.

Finally, amphibians had ____
____.

Additional notes about Evolution of Amphibians: ______________________
____
____

EVOLUTION OF REPTILES

Key Idea: The major evolutionary innovations that first appeared in reptiles include ______________, ______________ skin and the ______________ egg.

An **amniotic egg** is ____
____

The word **diversity** means ____
____

Additional notes about Kinds of Dinosaurs: ______________________
____
____

Additional notes about Extinction of Dinosaurs: ______________________
____

Reading Check: Why are amphibians tied to water, despite their adaptations to life on land? ____
____

Use a dictionary to find the meanings of the word parts in the terms *theropod* and *sauropod.* ____

____

EVOLUTION OF BIRDS

Key Idea: Birds first evolved about ______________ million years ago from ______________________________. The first birds had a skeleton that ______________________________. The first fossil found to show the link between ______________ and birds was that of ______________.

Additional notes about Age of Birds: ______________________________

____

____

Reading Check: What characteristics of Archaeopteryx were similar to the characteristics of birds today? ______________________________

____

EVOLUTION OF MAMMALS

Key Idea: The first mammals appeared about ______________ million years ago, not long after dinosaurs appeared. Mammals are descendants of the ______________.

A **therapsid** is ____

____

Additional notes about Ice Age Mammals: ______________________________

____

____

Reading Check: What were early mammals like? ______________________________

____

Name ______________ Class ______________ Date ______________

Note-taking Workbook

Simple Invertebrates

Section: Sponges

CHARACTERISTICS OF SPONGES

Key Idea: Sponges as classified as animals because they are ______________, are ______________, have no ______________ and contain some specialized cells.

A **choanocyte** is ______________

An **amoebocyte** is ______________

Additional notes about Body Plan: ______________

Additional notes about Feeding: ______________

SPONGE REPRODUCTION

Key Idea: Sponges reproduce both asexually and sexually. Most sponges are ______________, which means they produce both ______________ and ______________.

Additional notes about Asexual Reproduction: ______________

Additional notes about Regeneration: ______________

Reading Check: Describe how a sponge can asexually reproduce.

____

____

Complete the process chart below that shows the process of sexual reproduction in sponges.

1: Sperm released → 2: → 3: → 4: → 5: Surface attachment, growth and development

GROUPS OF SPONGES

Key Idea: The modern sponges are classified according to the ______________ of the skeleton in their body wall.

A **spicule** is ____

____

A **spongin** is ____

____

Additional notes about Sponge Skeletons: ______________

____

____

Additional notes about Sponge Classification: ______________

____

____

Reading Check: Compare the three main types of sponge skeletons.

____

____

Name ______________________ Class ______________ Date ______________

Note-taking Workbook

Simple Invertebrates

Section: Cnidarians

CHARACTERISTICS OF CNIDARIANS

Key Idea: Cnidarians have two basic body forms, called the ______________ and the ______________.

Medusa is ______________________________

A **polyp** is ______________________________

A **cnidocyte** is ______________________________

A **nematocyst** is ______________________________

The **planula** is the ______________________________

Additional notes about Movement and Response: ______________________________

Additional notes about Feeding: ______________________________

Additional notes about Reproduction: ______________________________

Reading Check: Summarize the life cycle of *Obelia*. ______________________________

Use the words *cnidocyte* and *nematocyst* in a sentence. ______________________________

How are these two words related? ______________________________

GROUPS OF CNIDARIANS

Key Idea: The three main groups of cnidarians are ________________,

____________________, and ____________________.

Additional notes about Hydrozoans: ______________________________

____

____

Additional notes about Scyphozoans: ____________________________

____

____

Additional notes about Anthozoans: ______________________________

____

____

Reading Check: How is the foundation of a coral reef formed?

____

____

Note-taking Workbook

Simple Invertebrates

Section: Flatworms

CHARACTERISTICS OF FLATWORMS

Key Idea: Flatworms have ________________ symmetry, ________________ tissue layers, and ________________ - the concentration of nerve tissue at an animal's "head" end.

Additional notes about Body Plan: ______________________________

____

Additional notes about Movement and Response: ________________

____

Additional notes about Feeding: ______________________________

____

Additional notes about Respiration: ___________________________

____

Additional notes about Reproduction: _________________________

____

Reading Check: How does a flatworm obtain oxygen? ____________

____

GROUPS OF FLATWORMS

Key Idea: Three major groups of modern flatworms include ____________, most of which are free-living, and ________________ and ____________, which are parasitic.

A **proglottid** is ______________________________________

____

The word **range** means ________________________________

____

Additional notes about Turbellarians: ____________________

____

____

Additional notes about Tapeworms: ____________________

____

____

Additional notes about Flukes: ____________________

____

____

Reading Check: Why is a tegument important to endoparasites? ________

____

Complete the process chart below that shows the life cycle of a fluke.

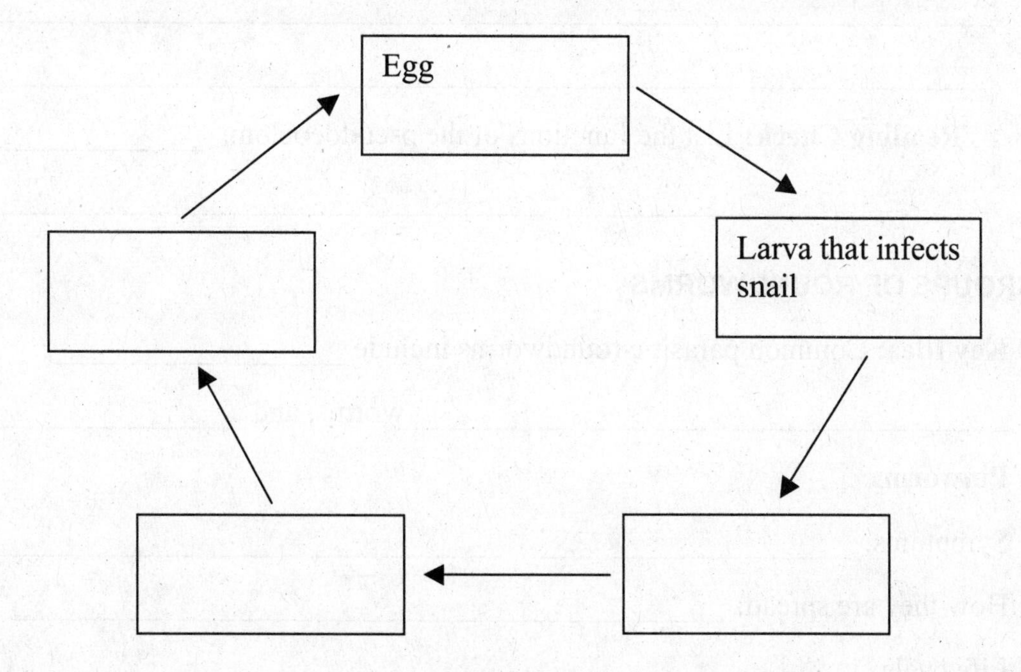

Name ______________________ Class ______________ Date ______________

Note-taking Workbook

Simple Invertebrates

Section: Roundworms

CHARACTERISTICS OF ROUNDWORMS

Key Idea: Roundworms have ________________ embryonic tissue layers, a ________________, and a digestive system with separate openings for feeding and waste elimination.

A **pseudocoelom** is ____

____

Additional notes about Body Plan: ______________________________

____

Additional notes about Digestion, Circulation, and Reproduction: ______

____

____

Reading Check: List the functions of the pseudocoelom. ____________

____

GROUPS OF ROUNDWORMS

Key Idea: Common parasitic roundworms include ________________, ________________, ________________ worms, and ________________.

Pinworms

Symptoms: ____

How they are spread: ______________________________________

Life cycle: ____

Hookworms:

Symptoms: ____

How they are spread: ______________________________________

Life cycle: ____

Filarial Worms:

Symptoms: ____

How they are spread: ____________________________________

Life cycle: ____

Ascarids

Symptoms: ____

How they are spread: ____________________________________

Life cycle: ____

Reading Check: Compare ascarids with hookworms. ____________

____

FIGHTING PARASITES

Key Idea: To prevent parasitic infections, people who live in or travel to places where parasites are common should ____________________

____

____.

The word **inadequate** means ____________________

____

Additional notes about Fighting Parasites: ____________

____

____

Name ______________________ Class ______________ Date ______________

Note-taking Workbook

Mollusks and Annelids

Section: Mollusks

CHARACTERISTICS OF MOLLUSKS

Key Idea: Mollusks are soft-bodied ______________ that have a three-part body plan. Mollusks also have ______________ symmetry, and most mollusks have a ______________.

Additional notes about Characteristics Mollusks: ______________

____

MOLLUSK BODY PLAN AND ORGAN SYSTEMS

Key Idea: The three parts that make up the basic mollusk body plan are the ______________ mass, the ______________, and the ______________.

Visceral mass is the ______________

____

A **mantle** is ______________

____

A **foot** is ______________

____

A **radula** is ______________

____

A **trochophore** is ______________

____

Additional notes about Feeding and Digestion: ______________

____

____

Additional notes about Excretion: ______________

____

Additional notes about Circulation: ______________________

____

Additional notes about Respiration: ______________________

____

Additional notes about Reproduction: ______________________

____

Reading Check: Describe a typical mollusk circulatory system. ________

____

MOLLUSK DIVERSITY

Key Idea: ________________, ________________, and ____________

share the same basic organ system and tissue layers, but have different ______

________________ and ________________.

A **siphon** is a ______________________________________

____

The word **equip** means ______________________________________

____

Additional notes about Mollusk Diversity: ______________________

____

____

Reading Check: What are the three major classes of mollusks? ________

____

Additional notes about Gastropods: ______________________

____

____

Reading Check: What are two feeding habits of gastropods? ________

____

Additional notes about Cephalopods: ______________________

____

____

Reading Check: How can some cephalopods escape predators?

____

____

Additional notes about Bivalves: ______________________

____

____

Summarize how mollusks are classified. ______________________

____

____

Name ______________________ Class ______________ Date ____________

Note-taking Workbook

Mollusks and Annelids

Section: Annelids

CHARACTERISTICS OF ANNELIDS

Key Idea: In addition to ______________, annelids are ____________ with highly specialized organ systems. Most annelids have external bristles called ______________.

A **seta** is ____

A **septum** is ____

The **cerebral ganglion** is ____

Additional notes about Segmentation: ____

Additional notes about Nervous System: ____

Additional notes about Reproduction and Development: ____

Reading Check: Describe the body plan of an annelid. ____

ANNELID DIVERSITY

Key Idea: Annelids are grouped into different classes based on the number of ______________ that they have and the presence or absence of ______________, which are flap-shaped appendages used for gas exchange and locomotion.

The word **region** means ____.

Additional notes about Marine Worms: ______________________________

Additional notes about Earthworms: ______________________________

Additional notes about Leeches: ______________________________

Reading Check: Explain how the chemicals that a leech's sucker secretes help a parasitic leech feed longer on its host. ______________________________

Use the comparison table below to compare different classes of annelids.

Marine worms (Polychaeta)	Earthworms (Oligochaeta)	Leeches (Hirudinea)

Name ______________________ Class ______________ Date ______________

Note-taking Workbook

Arthropods and Echinoderms

Section: Arthropods

ARTHROPOD CHARACTERISTICS

Key Idea: Arthropods are characterized by having a ______________ body, ______________ appendages, and a hard ______________.

A **thorax** is ______________________________

____

The **cephalothorax** is ______________________________

____

An **appendage** is ______________________________

____

A **trachea** is ______________________________

____

A **spiracle** is ______________________________

____

The **Malpighian tubule** is ______________________________

____

A **compound eye** is ______________________________

____

Additional notes about Segmented Body: ______________________________

____

Additional notes about Jointed Appendages: ______________________________

____

Additional notes about Exoskeleton: ______________________________

____

____

Reading Check: What does arthropod mean? ______________

____

Additional notes about Respiration and Circulation: ______________

____

Additional notes about Feeding, Digestion and Excretion: ______________

____

Additional notes about Compound Eyes: ______________

____

ARTHROPOD LIFE CYCLE

Key Idea: ______________ allows the arthropod's body to grow larger.

Molting is ______________

____

Additional notes about Molting: ______________

____

GROUPS OF ARTHROPODS

Key Idea: The four main arthropod groups are ______________,

______________, ______________, and ______________.

Additional notes about the Evolutionary Success of Arthropods: ______________

____

____

Reading Check: How does an arthropod shed its exoskeleton? ______________

____

Name ______________________ Class ______________ Date ______________

Note-taking Workbook

Arthropods and Echinoderms

Section: Arachnids and Crustaceans

ARACHNIDS AND THEIR RELATIVES

Key Idea: The ________________ are arthropods that have appendages called ________________, which are specialized for feeding.

A **chelicera** is ____

____

A **pedipalp** is ____

____

A **spinneret** is an ____

____

The word **modify** means ____

____

Additional notes about Spiders: ______________________________

____

Additional notes about Scorpions, Mites and Ticks: ________________

____

____

Additional notes about Horseshoe Crabs: ______________________

____

Reading Check: Name two types of spiders that are dangerous to humans.

____

____

CRUSTACEANS

Key Idea: Many crustaceans have a ________________ and an abdomen.

Like chelicerates, crustaceans have ________________ on their abdomen.

Unlike chelicerates, crustaceans have ________________ that are adapted for feeding and have two ________________. Crustaceans breathe by using ________________.

Additional notes about Terrestrial Crustaceans: ________________

____

____

Additional notes about Aquatic Crustaceans: ________________

____

____

Reading Check: Which structures do decapods use to swim and to reproduce? ________________________________

____

Classify arthropods into groups based on the types of appendages that the animals have. ________________________________

____

____

Name _______________ Class _______________ Date _______________

Note-taking Workbook

Arthropods and Echinoderms

Section: Insects

INSECT CHARACTERISTICS

Key Idea: Most insects share the same general body plan, specialized _______________ for feeding, a unique life style and the ability to _______________.

A **mandible** is _______________

____

Additional notes about Adaptations for Feeding: _______________

____

ADAPTATIONS FOR FLIGHT

Key Idea: Insects are adapted for flight by having a lightweight _______________, _______________ and strong _______________ to power flight.

Additional notes about Flying Insects: _______________

____

Reading Check: What are some advantages of flight? _______________

____

INSECT LIFE CYCLE

Key Idea: Insects have a unique life cycle compared with other arthropods. During development, a young insect undergoes _______________.

Metamorphosis is _______________

____

A **chrysalis** is _______________

____

A **pupa** is ____

____

Additional notes about Complete Metamorphosis: ________

____

____

Additional notes about Incomplete Metamorphosis: ________

____

____

Reading Check: What is metamorphosis? ____________

____

SOCIAL INSECTS

Key Idea: Social insects have elaborate social systems involving ________

____.

A **caste** is ____

____

Additional notes about Honeybees: ____________________

____

____

Additional notes about Termites: ____________________

____

____

CENTIPEDES AND MILIPEDES

Key Idea: ______________ include centipedes and millipedes.

Additional notes about Centipedes and Millipedes: ________

____

____

Name ______________________ Class ______________ Date ____________

Note-taking Workbook

Arthropods and Echinoderms

Section: Echinoderms

ECHINODERM CHARACTERISTICS

Key Idea: All adult echinoderms have an ______________ skeleton, ______________ symmetry, a ______________ system, and the ability to breathe through their skin.

An **ossicle** is ____

A **water-vascular system** is ____

A **tube foot** is ____

A **skin gill** is ____

Additional notes about Endoskeleton: ____

Reading Check: Why is an endoskeleton beneficial? ____

Additional notes about Five-Part Radial Symmetry: ____

Additional notes about Water-Vascular System: ____

Additional notes about Circulation and Respiration: ____________

____

____

ECHINODERM DIVERSITY

Key Idea: The living classes of echinoderms include ____________

____

____________.

Additional notes about Echinoderm Diversity: ____________

____

____

Reading Check: Which echinoderms disperse as larvae? ____________

____

____

Name ______________________ Class ______________ Date ______________

Note-taking Workbook

Fishes and Amphibians

Section: The Fish Body

CHARACTERISTICS OF FISHES

Key Idea: Fishes have ______________, ______________,

______________________________ and ______________.

Additional notes about Characteristics of Fishes: ______________

____

____

Reading Check: Why are fishes so diverse? ______________

____

MOVEMENT AND RESPONSE

Key Idea: Fishes have many important structures for swimming and sensing their underwater environment, including ______________

____.

The **swim bladder** is ______________________________

____

The **lateral line** is ______________________________

____

Additional notes about Endoskeleton: ______________

____

Additional notes about Fins and Swim Bladder: ______________

____

____

Additional notes about Sensory Organs: ______________

____

____

Reading Check: What is the function of the lateral line? ____________

____

RESPIRATION AND CIRCULATION

Key Idea: Fishes are able to obtain the oxygen they need from

________________.

A **gill** is a ______________________________________

____

A **gill slit** is a ___________________________________

____

Additional notes about Countercurrent Flow: ____________

____

____

Additional notes about Single-Loop Blood Circulation: ________

____

____

Reading Check: Describe how oxygen moves through a fish' body.

____

____

The prefix *counter* means "against." Write a definition in your own words for *countercurrent flow.* ______________________________

____

EXCRETION

Key Idea: Although the gills play a major role in maintaining a fish's

________________ and ________________ balance, another key element

is a pair of ________________.

A **kidney** is ____

____

The word **minimize** means ____

____

Additional notes about Salt and Water Balance: ______________

____

____

Additional notes about Kidneys: ______________

____

____

Reading Check: Compare marine and freshwater fish excretion. ______

____

____

REPRODUCTION

Key Idea: Most fishes reproduce sexually though ______________

______________.

Additional notes about Reproduction: ______________

____

____

Reading Check: Describe the process of spawning. ______________

____

Note-taking Workbook

Fishes and Amphibians

Section: Groups of Fishes

JAWLESS FISHES

Key Idea: Jawless fishes have skeletons made of ______________, a strong fibrous connective tissue. They retain their ______________ into adulthood. ______________ ______________ are the only modern vertebrates without a backbone.

Additional notes about Hagfishes: ______________________________

____

Additional notes about Lampreys: ______________________________

____

Reading Check: Compare the feeding methods of jawless fishes.

____

____

CARTILAGINOUS FISHES

Key Idea: Cartilaginous fishes have ______________________ and jaws. They also have skeletons made of cartilage strengthened by ______________ ______________________ (the material that makes up oyster shells).

The word **typically** means ______________________________

____

Additional notes about Sharks: ______________________________

____

Additional notes about Skates and Rays: ______________________________

____

Reading Check: List three characteristics of cartilaginous fishes.

____

____

BONY FISHES

Key Idea: Bony fishes have a strong endoskeleton made completely of ______________. Bony fishes also have structural adaptations, such as ____, that contribute to their success.

An **operculum** is a ____

____

A **teleost** is ____

____

Additional notes about Lateral Line: ______________________________

____

Additional notes about Gill Cover: ______________________________

____

Additional notes about Swim Bladder: ______________________________

____

Additional notes about Ray-finned Fishes: ______________________________

____

Additional notes about Lobe-finned Fishes: ______________________________

____

Reading Check: How does the fin of a teleost compare with the fin of a lo-finned fish? ____

____

Name ______________________ Class ______________ Date ______________

Note-taking Workbook

Fishes and Amphibians

Section: The Amphibian Body

CHARACTERISTICS OF AMPHIBIANS

Key Idea: Most of these amphibians share five key characteristics: ________

____

____.

Additional notes about Characteristics of Amphibians: ______________

____

Reading Check: Why do most amphibians live in moist habitats?

____

____

MOVEMENT AND RESPONSE

Key Idea: The sense of ______________ and ______________ are well developed in most amphibians. The primary sensory organs of amphibians are the ______________ and ______________.

The **tympanic membrane** is ______________________

____

The word **transmitted** means ______________________

____

Additional notes about Skeleton: ______________________

____

____

Additional notes about Sense Organs: ______________________

____

____

Reading Check: Describe how amphibians hear sounds. ________

____

RESPIRATION

Key Idea: In amphibians, the ______________, ______________,

________________________, and a partially divided

______________ work together to ensure that sufficient oxygen reaches

the body tissues.

A **lung** is ________________________________

____

Additional notes about Lungs: ________________________

____

____

Additional notes about Skin: ________________________

____

Reading Check: What structures do amphibians use to breathe?

____

____

CIRCULATION

Key Idea: The structure of the amphibian circulatory system – including the partially divided heart and double-loop circulation – allows oxygen to be delivered to the body more ______________ than in fishes.

The **septum** is ________________________________

____

The **pulmonary veins** are ________________________

____

Additional notes about Partially Divided Heart: ______________

____

____

Additional notes about Spiral Valve: ______________

____

____

Additional notes about Double-Loop Circulation: ______________

____

____

Reading Check: How is the circulatory system of amphibians different from the circulatory system of most fishes? ______________

____

Write two sentences that compare and two sentences that contrast fish and amphibian circulation. ______________

____

____

____

Name ______________________ Class ______________ Date ______________

Note-taking Workbook

Fishes and Amphibians

Section: Groups of Amphibians

SALAMANDERS

Key Idea: Salamanders have ____

____.

Additional notes about Salamanders: ______________________________

____

____

Reading Check: How do most salamanders reproduce?

____

____

CAECILIANS

Key Idea: Caecilians are a highly specialized group of burrowing amphibians with ______________, ______________________________ embedded in their skin.

Additional notes about Caecilians: ______________________________

____

____

FROGS AND TOADS

Key Idea: There are about ______________ species of frogs and toads, also known as ______________, which live in environments ranging from deserts to rain forests, valleys to mountains, and ponds to puddles.

A **tadpole** is ____

____

The word **cycle** means ______________________________

____

Additional notes about Frogs and Toads: ______________________________

____

____

____

Reading Check: How do the diets of adult and larval frogs differ?

____

____

Name _______________ Class _______________ Date _______________

Note-taking Workbook

Reptiles and Birds

Section: The Reptile Body

CHARACTERISTICS OF REPTILES

Key Idea: Modern reptiles have _______________, _______________, _______________, and reptiles lack _______________.

Additional notes about Characteristics of Reptiles: _______________

____

____

Reading Check: How are reptiles adapted to life on land? _______________

____

MOVEMENT AND REPSONSE

Key Idea: Reptiles have a strong _______________, _______________, _______________ and _______________.

These adaptations help reptiles _______________.

Jacobson's organ is _______________

____

Ectothermic is _______________

____

The word **determine** means _______________

____

Additional notes about Endoskeleton: _______________

____

Additional notes about Sensory Systems: _______________

____

Additional notes about Body Temperature Control: _______________

____

Reading Check: Why does being ectothermic limit reptiles' range?

____

____

RESPIRATION AND CIRCULATION

Key Idea: A reptile's lungs have a ______________ surface area, and a reptile's heart is almost completely divided into ______________ chambers.

Additional notes about Lungs: ____

____

Additional notes about Heart: ____

____

Reading Check: Compare reptilian and amphibian circulation.

____

____

REPRODUCTION

Key Idea: An ______________ egg contains both a water supply and a food supply and is key to ____.

Oviparous describes ____

____

Ovoviviparous describes ____

____

Additional notes about Amniotic Egg: ____

____

____

Reading Check: Summarize the structure of an amniotic egg. ________

____

Reptiles and Birds

Section: Groups of Reptiles

TURTLES AND TORTOISES

Key Idea: Turtles and tortoises have a hard shell that covers their body, and their ________________ is fused to the top of their shell.

The **carapace** is ____

____

The **plastron** is ____

____

Additional notes about Turtles and Tortoises: ________________

____

Reading Check: Describe the structure of a turtle's shell. ________

____

TUATARAS

Key Idea: Unlike most other reptiles, tuataras are more active at ________________ temperatures.

Additional notes about Tuataras: ________________________

____

____

CROCODILIANS

Key Idea: Unlike most other reptiles, crocodilians ________________ their young.

Additional notes about Crocodilians: ________________________

____

____

LIZARDS AND SNAKES

Key Idea: Because lizards and snakes share a common ancestor, they have many features in common, including

________________________________ and a jaw that is

________________ attached to the skull. This jaw allows the mouth to open wide enough to accommodate large prey.

A **constrictor** is a ____

____

The word **modify** means ______________________________________

____

Additional notes about Lizards: ______________________________

____

____

Additional notes about Snakes: ______________________________

____

____

Reading Check: Describe the ways that snakes kill their prey.

____

____

Note-taking Workbook

Reptiles and Birds

Section: The Bird Body

CHARACTERISTICS OF BIRDS

Key Idea: The key characteristics of modern birds include ______________

____

______________________________________.

Endothermic is ____

____

Additional notes about Body Temperature and Control: ______________

____

____

Reading Check: How s a bird able to maintain a high body temperature even in an artic environment? ______________________________

____

ADAPTATIONS OF BIRDS

Key Idea: ______________ and a ______________________ ______________ are critical to a bird's ability to fly.

A **contour feather** is ____

____

A **down feather** is ____

____

Additional notes about Feathers: ______________________________

____

Additional notes about Lightweight Skeleton: ____________________

____

Additional notes about Beaks and Feet: ____________________

____

____

Reading Check: What type of beak does a seed-eating bird have?

____

____

RESPIRATION AND CIRCULATION

Key Idea: Birds have respiratory and circulatory structures, such as ________ ________ and a ____________-chambered heart, that improve the efficiency of oxygen intake and oxygen delivery and allow birds to get the energy that they need.

The word **exceed** means ______________________________

____

Additional notes about Air Sacs: ______________________

____

Additional notes about Four-Chambered Heart: ____________

____

Reading Check: Describe the structure of a bird's heart.

____

____

Examine a figure showing a bird's air sacs during inhalation and exhalation. What do you think the prefixes *in-* and *ex-* mean? ____________

____

REPRODUCTION

Key Idea: Unlike most reptiles, most birds ________________ for their eggs and for their young.

Additional notes about Parental Care: ____________________________
____

Reading Check: Explain how different types of birds provide care for their young. ____
____

Name ______________________ Class ______________ Date ______________

Note-taking Workbook

Reptiles and Birds

Section: Groups of Birds

TERRESTRIAL BIRDS

Key Idea: Terrestrial birds have ______________ adapted to perching, hunting or running and ______________ adapted to eating fruits, seeds, insects or small animals.

A **talon** is ____

____

Additional notes about Perching Birds: ______________________

____

____

Additional notes about Birds of Prey: ______________________

____

____

Additional notes about Flightless Birds: ______________________

____

____

AQUATIC BIRDS

Key Idea: Aquatic birds have ______________ adapted to paddling or wading and ______________ adapted to eating aquatic organisms.

The word **maintain** means ______________________________

____

Additional notes about Diving Birds: ______________________

____

____

Additional notes about Water Birds: ______________________

____

____

Additional notes about Wading Birds: ______________________

____

____

Reading Check: What structures does a penguin use to swim?

____

____

Name ______________________ Class ______________ Date ______________

Note-taking Workbook

Mammals

Section: Characteristics of Mammals

KEY CHARACTERISTICS OF MAMMALS

Key Idea: Mammals are ________________, they have ________________ and ______________________________, and females produce milk in ________________ to nourish their young.

Additional notes about Key Characteristics of Mammals: ______________

____

____

Reading Check: What are three functions of hair? ________________

____

ENDOTHERMY

Key Idea: The ________________ and ________________ systems of mammals are adapted to ________________. They acquire and distribute oxygen more efficiently than the ectotherms do.

Additional notes about Respiratory System: ______________________

____

____

Additional notes about Circulatory System: ______________________

____

____

Reading Check: Why does a mammal need to eat more food than a reptile of a similar size does? ______________________________________

____

SPECIALIZED TEETH

Key Idea: Mammals have specialized teeth that reflect the differences in their

________________.

Additional notes about Types of Teeth: ______________________________

____

____

Reading Check: What type of tooth is used for stabbing and holding?

____

____

PARENTAL CARE

Key Idea: Unlike the young of other vertebrates, ________________ depend on their mother for a relatively long period of time. They receive __________ and other food, ________________, and ________________ from her. Mammals are unique in the way they ____________________________.

A **mammary gland** is ____

____

The **placenta** is ____

____

____

The **gestation period** is ____

____

Additional notes about Monotremes: ______________________________

____

____

Additional notes about Marsupials: ______________________________

____

____

Additional notes about Placental Mammals: ______________________

____

____

Reading Check: which group of mammals lay eggs? ______________

____

MOVEMENT AND RESPONSE

Key Idea: Mammals use various modes of ______________, including

______________________.

Echolocation is ______________________

____

The word **process** means ______________________

____

Additional notes about Movement and Response: ______________

____

____

Reading Check: What are three types of locomotion that mammals use?

____

____

Name ______________________ Class ______________ Date ____________

Note-taking Workbook

Mammals

Section: Groups of Mammals

MONOTREMES

Key Idea: Monotremes share more traits with ________________ than with other mammals. Monotremes are the only living mammals that ________________ ________________.

A **monotreme** is ____

____

Additional notes about Monotremes: ____

____

Reading Check: In what ways are monotremes more like reptiles than like other mammals? ____

____

MARSUPIALS

Key Idea: The females of most ________________ have a ________________, and their young spend most of their time ________________ ________________ while they ________________.

The word **similar** means ____

____

Additional notes about Marsupials: ____

____

____

Reading Check: What is an example of a marsupial whose ecological niche is similar to a deer's? ____

____

PLACENTAL MAMMALS

Key Idea: Placental animals make up nearly ______________ of all mammalian species. The young of placental mammals develop inside the female's ______________, where they are nourished by nutrients from her blood.

Additional notes about Placental Mammals: ______________

____

____

Reading Check: In general, which group of mammals has a longer gestation period: marsupials or placental mammals? ______________

____

Additional notes about Domestic Mammals: ______________

____

____

Reading Check: What are three ways in which humans use domestic mammals? ______________

____

Australia is the world's leading producer of wool, producing about 27% of the global total. If Australia produces 475,000 kg of wool per year, what is the global production of wool? ______________

____

Name ______________________ Class ______________ Date ____________

Note-taking Workbook

Mammals

Section: Evolution of Primates

CHARACTERISTICS OF PRIMATES

Key Idea: All primates have ______________________________ and most have ______________________________. The ____________________ placement of the eyes give primates overlapping, binocular vision.

A **primate** is ____

Additional notes about Characteristics of Primates: ______________

Reading Check: When did the first primates evolve? ______________

MODERN NONHMAN PRIMATE GROUPS

Key Idea: The three groups of primates alive today include ______________ and their relatives, ____________________, and ____________________ - monkeys, apes and humans.

Additional notes about Lemurs and Their Relatives: ______________

Additional notes about Tarsiers: ______________

Additional notes about Monkeys and Apes: ______________

EARLY HOMINIDS

Key Idea: Hominids differ from other primates in that we are ____________.

Our spines are ____-shaped, rather than ____-shaped. We have relatively short arms and a __________-shaped pelvis. Our thighs angle in under the body. Our spinal cord exits at the __________________ of the skull. And our canine teeth are ________________ than those of other primates.

A **hominid** is a ____
____
____

Additional notes about Bipedalism: ____________________
____
____

Additional notes about Studying Early Hominids: ____________
____
____

The Latin meaning of the species name *Homo erectus* is "upright man." Use a dictionary to find the meaning of the name *Homo sapiens.* ____________
____

THE PATH TO HUMANS

Key Idea: Many different hominid species lived over the past __________ years. And more than one species lived at the same time. Except for ________ ________________, all of these ancient hominid species died out.

The word **imply** means ________________________________

Additional notes about Earliest Hominids: ________________
____
____

Additional notes about Australopiths: ______________

____

____

Additional notes about *Paranthropus*: ______________

____

____

Additional notes about *Homo habilis:* ______________

____

____

Additional notes about *Homo ergaster* and *Homo erectus*: ______________

____

____

Reading Check: Which came first – a larger brain or bipedalism?

____

____

MODERN HUMANS

Key Idea: Studies of the genes of ______________ indicated that modern humans evolved in ______________ about ______________ years ago.

Additional notes about Neanderthals: ______________

____

____

Additional notes about *Homo sapiens*: ______________

____

____

Name ______________________ Class ______________ Date ______________

Note-taking Workbook

Animal Behavior

Section: The Nature of Behavior

QUESTIONS ABOUT BEHAVIOR

Key Idea: Scientists studying behavior ask two kinds of questions - ______________________ and ______________________.

Behavior is ____

Stimulus is ____

A **response** is ____

Additional notes about How vs. Why Questions: ____

Additional notes about How the Body Responds: ____

Reading Check: What is a response? ____

INFLUENCES ON BEHAVIOR

Key Idea: Behavior is controlled by both ______________ and ______________ factors.

Additional notes about Genes, the Environment, and Behavior: ____

Reading Check: What organ systems together produce behavior?

____

____

EVOLUTION OF BEHAVIOR

Key Idea: ______________________ favors traits that increase an individual's reproductive success.

Additional notes about Natural Selection and Behavior: __________

____

____

Reading Check: What traits are favored by natural selection? ________

____

The word *behavior* describes a phenotype, or trait of an organism. Use the word *behavior* in a sentence to illustrate its role as a trait, not just an action. ______

____

____

INNATE BEHAVIOR

Key Idea: Innate behaviors may be greatly influenced by

______________, but these behaviors are not necessarily fixed. Some innate behaviors may be modified by ______________.

Innate behavior is ________________________________

____

A **fixed action pattern** is ____________________________

____

Additional notes about Fixed Action Pattern: ______________

____

____

LEARNED BEHAVIOR

Key Idea: Learning can occur through ______________________________
______________________________.

Learning is ______________________________
____

Reasoning is ______________________________
____

Conditioning is ______________________________
____

Imprinting is ______________________________
____

The word **adaptive** means ______________________________
____

Additional notes about Habituation: ______________________________
____
____

Additional notes about Problem Solving: ______________________________
____
____

Reading Check: What is the value of habituation? ______________________________
____

Additional notes about Associative Learning: ______________________________
____
____
____
____

Additional notes about Sensitive Periods: ____________________

____

____

Reading Check: What are the benefits of imprinting behavior?

____

____

Name ______________________ Class ______________ Date ______________

Note-taking Workbook

Animal Behavior

Section: Classes of Behavior

SURVIVAL STRATEGIES

Key Idea: The survival of an individual depends on ____________________

____________________, such as ____________________, and ____________________

____________________.

Foraging is ____

____

Migration is ____

____

A **circadian rhythm** ____

____

Additional notes about Foraging Behavior: ____________________

____

____

Additional notes about Antipredator Behavior: ____________________

____

____

Additional notes about Cyclic Behavior: ____________________

____

____

MODES OF COMMUNICATION

Key Idea: Animals use ____________________ to influence the behavior of other animals. Many kinds of ____________________ can be used for this purpose.

Communication is ____

____

Additional notes about Sight: ______________________________

____

____

Reading Check: How has natural selection shaped signals? __________

____

Additional notes about Chemicals: __________________________

____

____

Additional notes about Touch: ______________________________

____

____

Additional notes about Sound: ______________________________

____

____

Additional notes about Language: ____________________________

____

____

Reading Check: Which signal is often used over large distances?

____

____

REPRODUCTIVE STRATEGIES

Key Idea: During a _________________ period, animals perform _________________ and _________________ behaviors. These behaviors, in different ways, maximize the reproductive success.

Territorial behavior is ______________________

____

The word **individual** means ______________________

____

Additional notes about Social Behavior: ______________________

____

____

Additional notes about Territorial Behavior: ______________________

____

____

Additional notes about Courtship Behavior: ______________________

____

____

Additional notes about Parental Care: ______________________

____

____

Additional notes about Cooperative Behavior: ______________________

____

____

Reading Check: How can parental care increase reproductive success?

____

____

Name ______________________ Class ______________ Date ______________

Note-taking Workbook

Skeletal, Muscular, and Integumentary Systems

Section: Body Organization

CELLS

Key Idea: ______________ cells are different from other cells of the body because they can ______________ and can become ______________________________.

A **stem cell** is ______________________________

Additional notes about Stem Cells: ______________________________

TISSUE TYPES

Key Idea: The human body contains four type of tissues: ______________________________.

Epithelial tissue is ______________________________

Nervous tissue is ______________________________

Connective tissue is ______________________________

Muscle tissue is ______________________________

Additional notes about Epithelial Tissue: ______________________________

Additional notes about Nervous Tissue: ______________________

____

____

Additional notes about Connective Tissue: ______________________

____

____

Additional notes about Muscle Tissue: ______________________

____

____

Reading Check: What are four types of tissues found in the body?

____

____

ORGANS AND ORGAN SYSTEMS

Key Idea: ________________ that work together form organs. Organs that work together form organ systems.

Additional notes about Organs and Organ Systems: ______________

____

____

HOMEOSTASIS

Key Idea: The body maintains homeostasis by ______________ and ______________ to changes in the internal environment.

Additional notes about Negative Feedback: ______________

____

Additional notes about Body Temperature: ______________

____

Name ______________________ Class ______________ Date ______________

Note-taking Workbook

Skeletal, Muscular, and Integumentary Systems

Section: The Skeletal System

THE SKELETON

Key Idea: The five important functions of the skeletal system are ______________

____

____.

Additional notes about Axial Skeleton: ______________________

____

Additional notes about Appendicular Skeleton: ________________

____

Reading Check: Which bones make up the pectoral girdle and which make up the pelvic girdle? ______________________________________

____

BONES

Key Idea: A typical bone is made up of four layers: ______________

____.

An **osteocyte** is ____

____

Bone marrow is ____

____

Leukemia is ____

The word **produce** means _______________________________________

Additional notes about Bone Structure: ______________________

____

____

Additional notes about Bone Growth: ______________________

____

Additional notes about Bone Injuries and Disorders: ______________________

____

Reading Check: How are osteocytes and osteoblasts different? ______________________

____

Complete this Venn diagram to help you remember the structure of bone.

Periosteum

Bone marrow

JOINTS

Key Idea: Movable joints are made up of ____________, ____________, and ____________.

A **joint** is ______________________

A **ligament** is ______________________

____

Additional notes about Types of Joints: ______________________

____

Additional notes about Joint Injuries and Disorders: ______________________

____

____

Reading Check: What is the difference between osteoarthritis and rheumatoid arthritis? ______________________

____

Name ______________________ Class ____________ Date ____________

Note-taking Workbook

Skeletal, Muscular, and Integumentary Systems

Section: The Muscular System

TYPES OF MUSCLES

Key Idea: The human body contains three types of muscle: ____________

____.

Additional notes about Types of Muscles: ____________________

____

MOVEMENT AND MUSCLE

Key Idea: Muscles switch between two processes to produce __________, depending on the level of ________________ and the presence of ________________.

A **tendon** is ____

____

A **flexor** is ____

____

An **extensor** ____

____

The word **flexible** means ____________________________________

____

Additional notes about Movement and Energy: ______________

____

____

Reading Check: What are the advantages and disadvantages of aerobic and anaerobic respiration? ______________________________

____

STRUCTURE OF MUSCLES

Key Idea: Skeletal muscle tissue is made of cells called ________________ ________________, which contain small cylinders called ________________. ________________ are made of ________________ linked end-to-end.

Muscle fiber is ________________

A **myofibril** is ________________

A **sarcomere** is ________________

Additional notes about Structure of Muscles: ________________

MUSCLE CONTRACTION

Key Idea: ________________ bind to ________________ filaments, ________________ filaments move inward, and ________________ shorten to cause muscle contraction.

A **myosin** is ________________

An **actin** is ________________

Steps of Muscle Contraction

Step 1 ________________

Step 2 ________________

Step 3 ________________

Step 4 ____

____

Additional notes about Muscle Contraction: ______________

____

____

Reading Check: How is ATP involved in muscle contraction? ________

____

The word part *myo-* means "muscle." The word part *sarco-* means "flesh." Why might *myo-* and *sarco-* both be used to refer to muscle? __________

____

EXERCISE AND MUSCLE

Key Idea: Exercise can affect ______________________________

______________. It can result in ____________ and __________.

And lack of exercise can cause ______________________________.

Additional notes about Strength, Speed, and Endurance: ________

____

____

Additional notes about Fatigue: ______________________________

____

____

Additional notes about Injury: ______________________________

____

____

Additional notes about Atrophy: ______________________________

____

____

Name ______________________ Class ______________ Date ______________

Note-taking Workbook

Skeletal, Muscular, and Integumentary Systems

Section: The Integumentary System

SKIN

Key Idea: The structure of skin includes ____.

The **epidermis** is ____

The **dermis** is ____

Subcutaneous tissue is ____

Additional notes about Epidermis: ____

Additional notes about Dermis: ____

Additional notes about Subcutaneous Tissue: ____

FUNCTIONS OF SKIN

Key Idea: The skin protects the body from ____.

Melanin is ____

Keratin is ____

Sebum is ____

____

The word **process** means ____

____

Additional notes about UV Protection: ____

____

Additional notes about Disease Prevention: ____

____

Additional notes about Temperature Regulation: ____

____

Additional notes about Waterproofing: ____

____

HAIR AND NAILS

Key Idea: Hair and nails are formed by ____

________________ and made of ________________.

Additional notes about Hair: ____

____

____

Additional notes about Nails: ____

____

____

Reading Check: Which structures produce hair and nails? ________________

____

DISORDERS OF SKIN

Key Idea: Skin disorders can be genetic, the result of ______________ or ______________, or a result of changes that occur within the body over time.

Additional notes about Acne: ______________________________

____

____

Additional notes about Skin Cancer: ______________________________

____

____

Additional notes about Other Disorders: ______________________________

____

____

Use what you have learned about skin disorders to solve the following analogy.

skin cancer : radiation :: ringworm : ______________

Additional notes about Blood Vessels: ______________________

____

____

Reading Check: How is blood pushed through the veins? ______________________

____

BLOOD

Key Idea: The key components of human blood are ______________________

______________________.

Plasma is ______________________

____

A **red blood cell** is ______________________

____

A **white blood cell** is ______________________

____

A **platelet** is ______________________

____

Additional notes about Blood: ______________________

____

____

Additional notes about Blood Types: ______________________

____

____

____

Reading Check: Which blood types can an A+ patient receive?

____

____

LYMPHATIC SYSTEM

Key Idea: The lymphatic system works with the cardiovascular system by collecting ________________ that leak our of ________________ and returning those ________________ to the cardiovascular system.

The **lymphatic system** is ____
____

Additional notes about Lymphatic System: ______________________
____

Reading Check: What are two ways in which the cardiovascular system and the lymphatic system are related? ______________________
____

Name ____________________ Class ______________ Date ____________

Circulatory and Respiratory Systems

Section: The Respiratory System

THE PATH OF AIR

Key Idea: Air enters the ________________ passage, then flows through the ________________, the ________________, the ________________, the ________________ tubes, and finally the ________________ into the ________________ of the lungs.

The **pharynx** is ____
____

The **larynx** is ____
____

The **trachea** is ____
____

The **bronchus** is ____
____

An **alveolus** is ____
____

The **diaphragm** is ____
____

Additional notes about The Path of Air: ________________________
____
____

BREATHING

Key Idea: When the diaphragm and rib muscles ________________, the chest cavity ________________ and air rushes ______. When these muscles relax, the chest cavity ____________________ and air rushes ________.

Additional notes about Inhalation and Exhalation: ______________

____

____

Additional notes about Breathing Rate: ______________

____

Reading Check: What causes air to enter the lungs? ______________

____

GAS EXCHANGE AND TRANSPORT

Key Idea: Oxygen is transported bound to ______________ inside

______________, and most carbon dioxide is carried as

______________ ions in the ______________.

The word **release** means ______________

____

Additional notes about Oxygen Exchange and Transport: ______________

____

____

Additional notes about Carbon Dioxide Exchange and Transport:

____

____

____

Reading Check: What signals hemoglobin to release O2?

____

____

RESPIRATORY DISEASES

Key Idea: Six common disease of the respiratory system are ____________

____.

Additional notes about Asthma: ____________________________________

____

____

Additional notes about Bronchitis: __________________________________

____

____

Additional notes about Pneumonia: __________________________________

____

____

Additional notes about Tuberculosis: ________________________________

____

____

Additional notes about Emphysema: _________________________________

____

____

Additional notes about Lung Cancer: ________________________________

____

____

Reading Check: Which lung diseases could be caused by smoking?

____

____

Name ______________________ Class ______________ Date ______________

Note-taking Workbook

Digestive and Excretory Systems

Section: Nutrition

FOOD AND ENERGY

Key Idea: Even when you are not moving, your body needs energy to

________________, ______________________________, and

__________. The more active you are, the more __________________ you

burn to release energy.

A **nutrient** is ____

____

A **calorie** is ____

____

Additional notes about Food and Energy: ______________________

____

FUEL FOR THE BODY

Key Idea: __________________, __________________, and ______________

provide most of the energy and building materials for the body.

Additional notes about Carbohydrates: ________________________

____

Additional notes about Proteins: ______________________________

____

Additional notes about Fats: __________________________________

____

Name a class to which the following members could belong: *glucose, sucrose, starch,* and *cellulose.* ____

OTHER ESSENTIAL NUTRIENTS

Key Idea: ________________, ________________, and ________________ contribute to many functions, such as regulating chemical reactions that release energy.

A **vitamin** is ____

A **mineral** is ____

Additional notes about Vitamins: ______________________________

Additional notes about Minerals: ______________________________

HEALTHY EATING HABITS

Key Idea: Good ________________ must be balanced with ________________________________ to maintain a healthy body.

The word **abnormal** means ______________________________

Additional notes about Body Mass Index: ______________________

Additional notes about Excess Body Fat: ______________________

Additional notes about Eating Disorders: ______________________

Reading Check: What is BMI? ______________________________

Name ______________________ Class ____________ Date ____________

Note-taking Workbook

Digestive and Excretory Systems

Section: Digestion

FROM FOOD TO NUTRIENTS

Key Idea: The ______________________ takes in food, breaks it down into ______________________, and gets rid of ______________________.

Digestion is ______________________

____

Additional notes about Chemical Digestion: ______________________

____

Additional notes about Mechanical Digestion: ______________________

____

STARTING DIGESTION

Key Idea: The ______________ and ______________ begin the process of digestion by breaking food into small particles.

An **esophagus** is a ______________________

____

Peristalsis is ______________________

____

Pepsin is ______________________

____

Additional notes about Mouth: ______________________

____

Additional notes about Stomach: ______________________

____

____

ABSORBING NUTRIENTS

Key Idea: Absorption of nutrients takes place primarily in the ____________ intestine, and is aided by secretions from the ______________ and ________________.

A **villus** is ____
____

Additional notes about Liver and Pancreas: ____________________
____
____

Additional notes about Small Intestine: ____________________
____
____

Reading Check: Where in the body does the digestion of fats begin?
____

REMOVING WASTE

Key Idea: All components of food that are not absorbed leave the body as ____________.

Additional notes about Large Intestine: ____________________
____
____

Additional notes about Water: ____________________
____
____

Reading Check: What are the contents of the large intestine? __________
____

Digestive and Excretory Systems

Section: Excretion

METABOLIC WASTES

Key Idea: By removing ______________ and ______________, excretion enables the body to maintain its __________ and ______ balance.

Excretion is ____

____

Urea is ____

____

Additional notes about Excretory Organs: ______________________

____

____

Additional notes about Water: ______________________________

____

____

CLEANING THE BLOOD

Key Idea: The ______________ filter wastes out of the blood and balance levels of molecules.

A **nephron** is ____

____

Additional notes about Kidney Structure: ______________________

____

____

Additional notes about Filtration: ____________________________

____

____

Additional notes about Reabsorption: ______________________________

____

____

Additional notes about Secretion: ______________________________

____

____

Reading Check: Why aren't blood cells in the filtrate? ______________

____

Complete the pattern puzzle below to help you remember how nephrons clean the blood.

Step 1: An arteriole enters the Bowman's capsule and splits into a bed of capillaries. Blood pressure forces fluid into the cup of the Bowman's capsule.

Step 2:

Step 3: The long, narrow tubes bend at their center to form a loop. Capillaries wrap around these tubules.

Step 4:

Step 5: The fluid passes out of the nephron through collecting ducts, where much of the water is removed.

URINARY EXCRETION

Key Idea: ________________ wastes are removed from the body through the formation and excretion of ________________.

Urine is ____

____

A **ureter** ________________________________

____

The **urinary bladder** is ________________________________

____

The **urethra** is ________________________________

____

Additional notes about Urinary Organs: ________________

____

____

Additional notes about Urination: ________________

____

____

Reading Check: What is a ureter? ________________

____

KIDNEY FAILURE

Key Idea: Kidneys are vital to maintaining ________________, so damage to the kidneys may eventually become life threatening.

Additional notes about Kidney Dialysis: ________________

____

____

Additional notes about Kidney Transplant: ________________

____

____

Reading Check: What are some common causes of kidney failure?

____

____

Name ______________________ Class ______________ Date ______________

Note-taking Workbook

The Body's Defenses

Section: Protecting Against Disease

PREVENTING ENTRY

Key Idea: ________________ and ______________________

form strong barriers that prevent pathogens from entering the body.

A **pathogen** is ______________________________________

____

A **mucous membrane** is ________________________________

____

Additional notes about Preventing Entry: ______________

____

____

Reading Check: How do physical barriers prevent pathogens from entering the body? ______________________________

NONSEPCIFIC IMMUNE RESPONSES

Key Idea: When pathogens break through the body's physical barriers, the body quickly responds with ________________________. These defenses are ________________, ________________, and ________________ of special proteins that kill or inhibit pathogens.

Inflammation is ____________________________________

____

Histamine is ______________________________________

____

The word part *path-* means "disease." The suffix *-gen* means "to bring forth." Write a sentence explaining how pathology and pathogens are related. _______

____

Additional notes about Fever: ______________________

____

____

Additional notes about Inflammation: ______________________

____

____

Additional notes about Protein Activation: ______________________

____

____

SPECIFIC IMMUNE RESPONSES

Key Idea: When a ______________ infects a cell, the body produces ______________ cells that specialize in detecting and destroying that specific pathogen.

A **macrophages** is ______________________

____

An **antigen** is ______________________

____

Additional notes about Antigen Display: ______________________

____

____

Additional notes about Two-Part Assault: ______________________

____

____

Reading Check: How does the body recognize "nonself" invaders?

____

____

Name ______________________ Class ______________ Date ____________

Note-taking Workbook

The Body's Defenses

Section: Eliminating Invaders

ACTIVATING A SPECIFIC IMMUNE RESPONSE

Key Idea: A specialized white blood cell called a ______________________ activates the immune system. These cells coordinate two responses: __________

____, and

____.

The **helper T cell** is ____

____

Additional notes about Helper T cells: ______________________

____

____

DESTROYING INFECTED CELLS

Key Idea: ______________________ destroy cells that have been infected by pathogens.

A **cytotoxic T cell** is a ____

____

Additional notes about Activating Cytotoxic T Cells: ______________

____

____

REMOVING PATHOGENS AT LARGE

Key Idea: The ______________________ response removes extracellular pathogens from the body and prevents further infection.

A **B cell** is ____

____

A **plasma cell** is ____

____

An **antibody** is ____

____

Additional notes about Activating B Cells: ______________________

____

____

Additional notes about Antibody Binding: ______________________

____

____

Reading Check: Which cells produce antibodies? ______________

____

Complete the pattern puzzle below to help you remember the steps of the immune response system.

Step 1: A virus, which displays viral antigens, can infect body cells

Step 2:

Step 3:

Step 4:

Step 5: B cells form specific plasma cells.

Step 6:

Step 7: Helper T ce start production of cytotoxic T cells th have the specific receptor.

Step 8:

LONG-TERM PROTECTION

Key Idea: After an immune response, ________________ cells continue to protect the body from pathogens the body had already encountered. An individual who recovers from an infectious disease becomes resistant to that particular pathogen.

A **memory cell** is a ______________________________

____

Immunity is ______________________________

____

A **vaccine** is ______________________________

____

Additional notes about Activating Memory Cells: ______________

____

____

Additional notes about Vaccination: ______________

____

____

Additional notes about How Pathogens Evade Immunity: ______________

____

____

Reading Check: What causes antigen shifting? ______________

____

Note-taking Workbook

The Body's Defenses

Section: Immune System Dysfunctions

ALLERGIES

Key Idea: An ______________________ is the immune system's excessive response to a normally harmless ________________.

An **allergy** is ____

____

An **allergen** is ____

____

Additional notes about Asthma: ______________________________

____

____

Additional notes about Other Allergies: ______________________

____

____

Reading Check: Why is asthma classified as an allergy?

____

____

AUTOIMMUNE DISEASES

Key Idea: In an autoimmune disease, the body launches an immune response, such that ________________ cells are attacked as if they were ________________.

An **autoimmune disease** is ______________________________________

____

Additional notes about Common Autoimmune Disorders: ____________

____

____

IMMUNE DEFICIENCY

Key Idea: When the immune system does not function, the body is unable to fight and survive infections by ________________ that do not cause any problems for a robust immune system.

AIDS is ____

____

HIV is ____

____

Additional notes about Innate Immune Deficiency: ____________

____

____

Additional notes about Immune Suppression: ________________

____

____

Additional notes about AIDS: ______________________________

____

____

Reading Check: Which immune cells does HIV destroy, and how does the virus destroy them? ________________________________

____

Create a mnemonic device to help you remember the types of immune system dysfunctions described in this section. ______________________

____

Name ______________________ Class ______________ Date ______________

Note-taking Workbook

Nervous System

Section: Structures of the Nervous System

CENTRAL NERVOUS SYSTEM

Key Idea: The central nervous system responses to ______________ and ______________ information.

A **neuron** is ____

____

The **central nervous system** is _______________________________

____

The **peripheral nervous system** is ____________________________

____

The **brain** is ____

____

The **cerebrum** is ____

____

The **brainstem** is ____

____

____

The **cerebellum** is ____

____

____

The **spinal cord** is ____

____

Additional notes about Brain: _________________________________

____

____

____

Additional notes about Spinal Cord: ____________________

____

____

____

Reading Check: What is a neuron? ____________________

____

PERIPHERAL NERVOUS SYSTEM

Key Idea: The peripheral nervous system contains ________________ and ______________________________ that carry information between the central nervous system and the rest of the body.

Additional notes about Autonomic Nervous System: ____________

____

____

Additional notes about Somatic Nervous System: ______________

____

THE SPINAL REFLEX

Key Idea: A spinal reflex is an involuntary ______________ triggered by sensory input and produced by neural circuitry limited to the ______________________________.

A **reflex** is ____________________________________

____

Additional notes about The Spinal Reflex: ____________________

____

____

Reading Check: What neurons are involved in a spinal reflex? ________

____

Note-taking Workbook

Nervous System

Section: Neurons and Nerve Impulses

OUR ELECTRICAL BODY

Key Idea: Electrical signals in the nervous system are caused by the movement of ______________ across the cell membrane of ______________.

Additional notes about Our Electrical Body: ______________

____

____

STRUCTURE OF NEURONS

Key Idea: A neuron's ______________ gather information from other cells, the cell body integrates this information, and the ______________ sends the information to other cells.

A **dendrite** is a ______________________________

____

An **axon** is ______________________________

____

A **nerve** is ______________________________

____

Additional notes about Structure of Neurons: ______________

____

____

Reading Check: What part of the cell sends electrical signals?

____

____

GENERATING A NERVE IMPULSE

Key Idea: All nerve impulses begin when the resting state of a ___________ is changed by a signal from another ________________ or from the environment.

Membrane potential is ____

____

____

Action potential is ____

____

____

The word **distribution** means ________________________________

____

Additional notes about Resting Potential: ____________________

____

____

Additional notes about Action Potential: ____________________

____

____

Additional notes about Threshold: __________________________

____

Reading Check: What structure keeps a neuron in its resting state?

____

____

Write a short paragraph describing threshold in your own words. __________

____

____

COMMUNICATION BETWEEN NEURONS

Key Idea: Neurons communicate with other cells at specialized junctions called ________________. Chemicals are released at ________________ and can stimulate nearby cells.

A **synapse** is ____
____

A **neurotransmitter** is ____
____

Additional notes about Neurotransmitter Release: ____________
____
____

Additional notes about Neurotransmitter Action: ____________
____
____

Additional notes about Neurotransmitter Removal: ____________
____
____

Reading Check: How are neurotransmitters released? ____________
____

Name ______________________ Class ______________ Date ______________

Note-taking Workbook

Nervous System

Section: Sensory Systems

PERCEPTION OF STIMULI

Key Idea: Specialized neurons called ______________________ detect sensory stimuli and convert them to ______________. These signals then can be interpreted by the brain.

Additional notes about Perception of Stimuli: ______________________

____

Reading Check: What types of cells are sensory receptors? __________

____

SENSORY RECEPTORS

Key Idea: The major classes of sensory receptors are ______________, ______________, ______________, ______________, and ______________.

A **sensory receptor** is ______________________

____

The **retina** is ______________________

____

The **cochlea** is ______________________

____

The **semicircular canal** is ______________________

____

A **taste bud** is ______________________

____

The word **generate** means ______________________

____

Additional notes about Vision: ______________________

____

____

Additional notes about Hearing and Balance: ______________________

____

____

Additional notes about Taste and Smell: ______________________

____

Additional notes about Touch and Other Sense: ______________________

____

____

Reading Check: How do hair cells detect sound waves? ______________________

____

Complete the spider map below about types of sensory receptor.

sensory receptors
chemoreceptors
photoreceptors
mechanoreceptors
pain receptors
thermoreceptors

PROCESSING OF SENSORY INFORMATION

Key Idea: Specialized regions of the ______________________

detect different ______________ information.

Additional notes about Processing of Sensory Information: ______________________

____

____

Name ______________________ Class ______________ Date ______________

Note-Taking Workbook

Nervous System

Section: Nervous System Dysfunction

PSYCHOACTIVE DRUGS

Key Idea: Many psychoactive drugs produce ______________________ and ______________ when abused.

A **psychoactive drug** is ______________________

____

A **depressant** is ______________________

____

A **stimulant** is ______________________

____

Reading Check: How do depressants cause death when they are abused?

____

____

Additional notes about Nicotine: ______________________

____

____

Additional notes about Alcohol: ______________________

____

____

Additional notes about Marijuana: ______________________

____

____

Additional notes about Narcotics: ______________________

____

____

Reading Check: Why are opiates used in pain relievers? ____________

____

NEURAL CHANGES

Key Idea: ______________ occurs when repeated use of a drug alters the normal functioning of ______________ and ______________.

An **addiction** is ____

____

A **tolerance** is ____

____

Withdrawal is ____

____

Additional notes about Neural Changes: ______________

____

____

Reading Check: How does addiction change a neuron? ____________

____

____

NERVOUS SYSTEM DISORDERS

Key Idea: Damage to the nervous system can occur as a result of the ________

____

______________.

The word **induce** means ____

____

Additional notes about Multiple Sclerosis: ______________________

____

____

Additional notes about Meningitis: ______________________

____

____

Additional notes about Traumatic Injury: ______________________

____

____

Reading Check: How can the majority of traumatic injuries be prevented?

____

____

Name ______________ Class ______________ Date ______________

Note-taking Workbook

The Endocrine System

Section: Hormones

WHAT IS THE ENDOCRINE SYSTEM?

Key Idea: The endocrine system regulates ______________________________

____

____.

A **hormone** is ______________________________

____

Additional notes about What is the Endocrine System?: ______________

____

____

WHERE ARE HORMONES MADE?

Key Idea: ______________ glands and ______________ tissue produce and release hormones.

An **endocrine gland** is ______________________________

____

Additional notes about Endocrine and Exocrine Glands: ______________

____

____

Additional notes about Endocrine Tissues: ______________________

____

____

Reading Check: What is the difference between endocrine glands, endocrine tissues, and exocrine glands? ______________________

____

HORMONES AND RECEPTORS

Key Idea: Hormones can affect only cells that have ______________ ______________ that match the hormones.

A **target cell** is ____

____

Additional notes about Types of Hormones: ______________

____

____

Additional notes about Hormone Receptors: ______________

____

____

Reading Check: How do amino acid-based and cholesterol-based hormones differ? ____

____

HORMONE FUNCTION

Key Idea: Amino acid-based hormones bind to the surface of ______________ ________ and cause changes by a second messenger system. Steroid hormones enter ______________ ________ and cause changes by direct gene activation.

A **second messenger** is ____

____

Steps of the Second Messenger System

Step 1: Glucagon binds to a receptor on the surface of a cell.

Step 2: ____

____

Step 3: Cyclic AMP starts a cascade of enzyme activation.

Step 4: ____

____

Steps of Direct Gene Activation

Step 1: Cortisol enters the cell and binds to a receptor.

Step 2: ______________________________

Step 3: ______________________________

Step 4: ______________________________

Step 5: The activities of the cell change.

Additional notes about The Second Messenger System: ______________________________

____

Additional notes about Direct Gene Activation: ______________________________

____

Reading Check: How does glucagons cause the production of glucose in a cell? ______________________________

____

Write cause-and-effect statements that describe how hormones affect target cells by the second messenger system. ______________________________

____

____

CONTROL OF HORMONE LEVELS

Key Idea: The body uses ______________________________

______________________________ to regulate the levels of hormones and their products.

A **feedback mechanism** is ______________________________

____

An **antagonistic hormone** is ______________________________

____

____

The word **specific** means ______________________

____

Additional notes about Negative Feedback: ______________________

____

____

Additional notes about Positive Feedback: ______________________

____

____

Additional notes about Antagonistic Hormones: ______________________

____

____

Additional notes about Endocrine System Disorders: ______________________

____

____

Reading Check: What is the difference between negative and positive feedback mechanisms? ______________________

____

Name ______________________ Class ______________ Date ______________

Note-taking Workbook

The Endocrine System

Section: Major Endocrine Glands

CONTROLLING THE ENDOCRINE SYSTEM

Key Idea: The ________________ and the ______________________

work together to control the functions of the endocrine system.

The word **summarized** means ______________________________

____

Additional notes about The Hypothalamus: ____________________

____

____

Additional notes about The Pituitary Gland: ____________________

____

____

Reading Check: Which hormones control the endocrine system?

____

____

REGULATING METABOLISM

Key Idea: The ______________________________________

____

regulate the body's metabolic processes.

Additional notes about The Thyroid Gland: ____________________

____

____

Additional notes about The Parathyroid Glands: ________________

____

____

Additional notes about The Pancreas: ______________________

____

____

Additional notes about The Pineal Gland: ______________________

____

____

RESPONDING TO STRESS

Key Idea: The ______________________ regulates short-term responses to stress. The ______________________ regulates long-term responses to stress.

Epinephrine is a ______________________

____

Norepinephrine is a ______________________

____

Additional notes about Short-Term Responses: ______________________

____

____

Additional notes about Long-Term Responses: ______________________

____

____

Reading Check: Which hormones are part of the body's long-term response to stress? ______________________

____

REGULATING REPRODUCTION

Key Idea: Reproduction is regulated by the ________________, which are released by the pituitary gland, and by sex hormones, which are released by the ________________.

An **androgen** is a ____

____

Estrogen is ____

____

Progesterone is ____

____

Additional notes about Gonadotropins: ________________

____

____

Additional notes about Gonads: ____________________

____

____

Name ____________________ Class ______________ Date ______________

Note-taking Workbook

Reproduction and Development

Section: The Male Reproductive System

SPERM PRODUCTION

Key Idea: The male reproductive system has two ______________ that produce the male gametes, ______________, and the primary sex hormone, ______________.

The **testis** is ____

____

A **seminiferous tubule** is ____

____

Additional notes about Seminiferous Tubules: ______________

____

____

Reading Check: Describe how hormones regulate sperm production.

____

____

SPERM'S PATH THROUGH THE BODY

Key Idea: The ______________________ carries sperm into the ______________. Sperm leave the body by passing through the ______________, the same duct through which urine exits the body.

The **epididymis** is ____

____

The **vas deferens** is ____

____

Additional notes about Structure of Mature Sperm: ______________

____

Reading Check: Why does a sperm cell need mitochondria?

____

____

SEMEN

Key Idea: As sperm cells move into the urethra, they mix with fluids secreted by the ______________________, the ______________________ and the ______________________.

The **prostate gland** is ______________________

____

Semen is ______________________

____

Additional notes about Semen: ______________________

____

____

SPERM DELIVERY

Key Idea: After semen is deposited in the female reproductive system, sperm swim until they encounter an ______________ or until they die.

The **penis** is ______________________

____

Additional notes about Sperm Delivery: ______________________

____

____

Reading Check: Explain why a high sperm count is necessary for fertilization. ______________________

____

Name ____________________ Class ____________ Date ____________

Note-taking Workbook

Reproduction and Development

Section: Human Development

FERTILIZATION

Key Idea: During fertilization, a __________ cell penetrates an __________ by releasing enzymes from the tip of its head. These enzymes break down the jellylike outer layers of the __________. The head of the __________ enters the __________, and the nuclei of the __________ and __________ fuse together.

An **embryo** is ______________________________

____

Implantation is ______________________________

____

Additional notes about Cleavage and Implantation: ______________________________

____

____

FIRST TRIMESTER

Key Idea: All of the embryo's ______________________________, as well as the supportive __________ that feed and protect the embryo, develop during the first trimester of pregnancy.

A **fetus** is ______________________________

____

Additional notes about Supportive Membranes: ______________________________

____

Additional notes about Embryonic Development: ______________________________

____

____

Additional notes about Fetal Development: ______________

____

____

Reading Check: During which trimester do the embryo's organs begin to form? ______________________

____

SECOND AND THIRD TRIMESTERS

Key Idea: By the end of the third trimester, the fetus is able to __________ outside the mother's body. The fetus leaves the mother's body in a process called ______________, which usually lasts several hours.

Additional notes about Second and Third Trimesters: __________

____

____

Reading Check: How does the fetus change in the third trimester?

____

____

Name ______________________ Class ____________ Date ____________

Note-taking Workbook

Reproduction and Development

Section: Sexually Transmitted Infections

STI TRANSMISSION AND TREATMENT

Key Idea: ________________ is the only sure way to protect yourself from contracting an STI.

Additional notes about STI Transmission and Treatment: ____________

____

____

Reading Check: How are different kinds of STIs treated?

____

____

COMMON STIS

Key Idea: The most common STIs include ____________________

____

Genital herpes is ______________________________

____

Pelvic inflammatory disease is ____________________

____

The word **available** means ________________________

____

Additional notes about Genital Herpes: ____________________

____

____

Additional notes about Genital HPV: ____________________

____

____

Additional notes about AIDS: ________________________________

____

____

Additional notes about Hepatitis B: ____________________________

____

____

Additional notes about Chlamydia and Gonorrhea: ________________

____

____

Additional notes about Syphilis: ________________________________

____

____

Name ______________________ Class ______________ Date ______________

Note-taking Workbook

Forensic Science

Section: Introduction to Forensics

WHAT IS FORENSIC SCIENCE?

Key Idea: The two major duties of a forensic scientist are ________________ and ________________.

The word **portrayed** means ________________

Additional notes about Analyze Evidence: ________________

Additional notes about Testify in Court: ________________

Reading Check: What are the two meanings of the word *identity* as it relates to forensic science? ________________

Additional notes about More Than Crime Scenes: ________________

Reading Check: List at least 10 things that forensic scientists do in addition to investigating crime scenes. ________________

Search the text to find examples of jobs that forensic scientists perform. Can you find more than 15 examples? ________________

TOOLS OF FORENSIC SCIENCE

Key Idea: Five basic types of tools used by forensic scientists are __________

__________________, __________________, __________________, and

__________________.

A **chromatograph** is a ______________________________

____

A **spectrometer** is a tool that ______________________________

____

Additional notes about Crime Scene Chemistry: ______________

____

____

Additional notes about Looking at Details: ______________

____

____

Additional notes about Identifying Substances: ______________

____

____

Additional notes about Organizing Information: ______________

____

____

Reading Check: What are three types of microscopes used by forensic scientists for examining evidence? ______________________

____

Name ______________________ Class ______________ Date ______________

Note-taking Workbook

Forensic Science

Section: Inside a Crime Lab

IDENTIFICATION AND DNA

Key Idea: Two unique characteristics that are used to identify people are

________________ and ________________.

The word **analyzed** means ________________________________

____

Additional notes about Identification: ________________________

____

____

Reading Check: Where are friction ridges found on the body? ________

____

Additional notes about DNA Analysis: ________________________

____

____

TRACE EVIDENCE

Key Idea: The ____________ major types of trace evidence are ________,

____________, ____________, ____________, and ____________.

Additional notes about Hair and Fiber: ________________________

____

Additional notes about Glass: ________________________________

____

Additional notes about Paint: ________________________________

____

Additional notes about Soil: ________________________________

____

Additional notes about Pollen and Other Clues: ______________

____

Complete the spider map below to help you organize details that describe various types of trace evidence.

Types of Trace Evidence

- paint
- soil
- pollen and other
- hair and fiber
- glass

FIREARMS AND TOOLMARKS

Key Idea: Firearms and toolmarks examiners study evidence from __________, __________, __________, __________, and __________, and perform __________.

Ballistics is ____

____

Additional notes about Firearms: ______________

____

Additional notes about Toolmarks: ______________

____

Reading Check: Which science deals with projectiles? __________

DRUGS, ALCOHOL, AND TOXICOLOGY

Key Idea: Forensic toxicologists analyze __________, __________, and __________ to detect the presence of chemicals and to determine whether a chemical was __________ or __________.

Toxicology is ____

____

Additional notes about Forensic Toxicology: ________________

____

____

Additional notes about Forensic Alcohol: ________________

____

____

Additional notes about Illegal Drugs: ________________

____

____

Reading Check: What is the difference between antemortem toxicologists and postmortem toxicologists? ________________

____

PATHOLOGY

Key Idea: Forensic pathologists perform ________________ to determine

____.

Pathology is ____

An **autopsy** is an ____

____

Additional notes about Branches of Pathology: ________________

____

____

Additional notes about Performing an Autopsy: ________________

____

____

Reading Check: What does a forensic pathologist look for during an autopsy? ____

ANTHROPOLOGY AND ENTOMOLOGY

Key Idea: Anthropology can help determine the identity of ________________ ________________________________. Anthropology and entomology can be used to ____.

Additional notes about Forensic Anthropology: ________________________

____

____

Additional notes about Forensic Entomology: ________________________

____

____

Reading Check: What are four things that you can learn about a person from a skeleton? ____

____

Name ____________________ Class ____________ Date ____________

Note-taking Workbook

Forensic Science

Section: Forensic Science in Action

AT THE CRIME SCENE

Key Idea: At a crime scene, ______________ secure a perimeter and interview witnesses; crime scene investigators ______________, ______________, ______________, and ______________ evidence.

Additional notes about Secure and Search: ______________

____

____

Additional notes about Photograph and Document: ______________

____

____

Additional notes about Collect and Analyze: ______________

____

____

Reading Check: What are three types of analysis that must be done at the crime scene? ______________

____

TIME OF DEATH

Key Idea: ______________, ______________, ______________, and the stages of ______________ are used to estimate the time at which a person died.

Rigor mortis is the ______________

____

Livor mortis is the ______________

____

Algor mortis is ____________________________________

Additional notes about Rigor Mortis: ____________________________________

Additional notes about Livor Mortis: ____________________________________

Additional notes about Algor Mortis: ____________________________________

Additional notes about Decomposition: ____________________________________

Complete the graph below to show the progress of rigor mortis, livor mortis, and algor mortis over the first 24 hours following death.

Progress of post-mortem event						Temperature (F)
maximum						97
						94
obvious						91
						84
beginning						80
	1	3	5	10	12	

Time post-mortem (h)

CAUSE, MECHANISM, AND MANNER OF DEATH

Key Idea: The ______________, __________________, and ______________ describe how a person died.

Additional notes about Cause: ____

____

____

Additional notes about Mechanism: ________________________________

____

____

Additional notes about Manner: ___________________________________

____

____

VICTIM AND PERPETRATOR

Key Idea: The identities of the ________________ and of the ______________ must be discovered in order to ____.

The word **establish** means to ______________________________________

____

Additional notes about Identity of the Victim: _____________________

____

____

Additional notes about Identity of the Perpetrator: __________________

____

____